Keep Your Eye on the Ball

Curve Balls, Knuckleballs,
and Fallacies of
Baseball

Keep Your Eye on the Ball

Curve Balls, Knuckleballs, and Fallacies of Baseball

Revised and Updated

Robert G. Watts

A. Terry Bahill

W. H. Freeman and Company
New York

Interior Design by Blake Logan

Library of Congress Cataloging-in-Publication Data

Watts, Robert G.
  Keep your eye on the ball / Robert G. Watts, A. Terry Bahill.—[Revised and Updated ed.]
    p. cm.
  Includes bibliographical references and index.
  ISBN 0-7167-3717-5
  1. Physics. 2. Baseball. 3. Force and energy. I. Bahill, Terry. II. Title.

QC 26.W38 2000
796.357'01'53—dc21   99-059352

Printed in the United States of America

First printing 2000

W. H. Freeman and Company
41 Madison Avenue, New York, NY 10010
Houndmills, Basingstoke RG21 6XS, England

*We dedicate this second edition to the memory of Joe DiMaggio.
Where have you gone, indeed.*

Dedication from the first edition

To B, who hit us fly balls when he could, to Don, who caught them all
gracefully and who hit me like a country cousin, and to Adam, who was
developing a fair curve ball when he took up cycling instead.

RGW

To Alex, Zach, and all the people who swung bats in our
experiments.

ATB

Contents

Preface

"KEEP YOUR EYE ON THE BALL." That is good advice as long as you realize that it cannot be done. In this book, we explain why the batter cannot keep his eye on the ball. We also explain why the curve ball curves. We discuss the knuckleball: the reason it is so effective, even though it is a slow pitch, is that its trajectory really is erratic; it can make an abrupt shift in the last 5 feet when the batter has already taken his eye off the ball. We also discuss collisions between ball and bat and history and statistics of the sport. We do all of this using scientific and engineering tools and techniques, but we have tried to write this book so that it can be understood by everyone.

In 1982, three scientists, Jim Walton of General Motors, Charlie Miller of MIT and one of us (Bob Watts) had the privilege of photographing and analyzing a series of pitches thrown by Scott McGregor, a pitcher for the Baltimore Orioles, and Ray Miller, then the pitching coach for the Orioles. Miller, especially, was quite curious about what we were doing, and we had some lively, entertaining, educational discussions. It has been our experience that baseball players, coaches, and fans alike are eager to discuss the science of baseball. They tend to be intelligent people who have some ideas of their own and are anxious to learn more about how science relates to their trade. These discussions spawned this book.

The book is written for the inquiring layperson or the fan who wonders why curve balls curve (and if they do, how and why), how knuckleballs and spitballs move, and how pitching a ball with backspin or overspin affects the batted ball. Although there are self-styled pundits of the game who claim to know the answers to these questions, we take the approach that only by rigorously applying the laws of physics, psychology and physiology can these matters be resolved once and for all. Thus many controversial claims about the game are addressed and, we hope, resolved by this book.

The mere mention of the word *physics* may have a chilling effect on some readers. Therefore, we have tried to simplify the scientific discussions by using a minimal number of laws and equations. We have put much of the information in the form of tables and graphs, and we made the figure captions descriptive and complete to guide the readers through most of our explanations of the science of baseball. When we felt a more scientific background might be helpful, we provided boxes within the text to avoid interrupting the narrative of the discussion. We hope that, with a little concentration and adequate motivation, you will easily be able to understand the more technical aspects of analyzing the game of baseball. We hope that high school mathematics and physics teachers will use this book to enliven their lectures. We have also written the book so that baseball fans with little appreciation for mathematics and physics can skip the equations.

We begin the book with a little bit of history about the origin and development of baseball in chapter 1. In chapter 2, we focus our historical discussion on the curve ball and begin our scientific discussion by explaining in general terms why the curve ball curves. In chapter 3, using more principles from physics, we explain the origin of the forces that cause the ball to curve. Then, in chapter 4, using greater mathematical detail, we show how large the forces are and how far they can make a ball curve. We describe several scientific experiments that were conducted to measure the forces on spinning and nonspinning balls. The results of these experiments, together with the equations developed in earlier chapters, are then used to calculate the deflections actually experienced by curve balls, knuckleballs and other pitches.

In chapter 5, we discuss bat-ball collisions and show how the ideal bat weight can be determined for each individual. We show that some batters could profit from corking their bats and others could be hurt. In chapter 6, we discuss the trajectory of a fly ball and explain the advantages and disadvantages of putting backspin or topspin on the batted ball.

In chapter 7, we show that batters cannot keep their eye on the ball. They can track it until it is about 5 feet in front of the plate, and then they fall behind because it is moving too fast. These physiological limitations cause batters to adopt tracking strategies that can produce the perceptual illusion of the rising fastball.

Many physiological parameters can be used to predict whether a particular ballplayer will be successful. For example, it is helpful if the player is big, strong and fast. In chapter 8, we discuss many other quantitative parameters that could be used. In particular we show that good batters quickly learn to track novel visual targets.

Finally, a chapter on statistics raises some intriguing questions about how to compare the exploits of baseball stars who played during different eras.

Many people have given us advice and assistance while we were doing the research and writing this book. We would like to thank all of them, in particular Jerry Kindall and Jerry Stitt with the University of Arizona baseball team; Al Rosen, Brett Butler and Chris Speier of the San Francisco Giants; Branch Rickey, Brian Harper and Joe Orslak of the Pittsburgh Pirates Baseball Club; Mike Candrea of the University of Arizona softball team; Steve Bahill of Tata's Machine Shop; Bill Karnavas of IBM Transarc Lab; Pete Brancazio of Brooklyn College; and Cliff Frohlich of the University of Texas.

The Game of Baseball:

A Little History

Social critic Jacques Barzun wrote that anyone who wants to know the heart and mind of America had better know baseball. Why is baseball such a typically American game? Was it, as some folk history would have us believe, invented by the American Abner Doubleday on a field in Cooperstown, New York, in 1839? Or, rather, is it an outgrowth of an English game called *rounders,* whose rules were printed in *The Boy's and Girl's Book of Sports,*[1] written by William Clarke and originally published in London in 1829?

The claim that Abner Doubleday invented baseball began in 1889 at Delmonico's Restaurant in New York City. A group of several hundred people had gathered to honor a squad of professional baseball players who, under the leadership of Albert G. Spalding, president of the Chicago Baseball Club, had just returned from a world tour. At the time Spalding was spreading the gospel that baseball was America's national game. One speaker at the banquet, Abraham C. Mills, the fourth president of the National League, stated that he wanted it understood that patriotism

1

and research had established that the game of baseball was American in origin.

Americans of the time would have found it difficult to admit that their favorite pastime was not American in origin. Many, however, including Henry Chadwick, one of the first great baseball sportswriters, had claimed before the Delmonico's banquet—and still claimed after it—that American baseball had originated from the English game of rounders. Spalding, on the other hand, championed the American-origin theory. He finally met Chadwick's challenge by calling for a "blue ribbon commission," consisting of seven "men of high repute." The commission was a handpicked crew made up of men who were at one time or another prominent in organized American baseball. One of these men was Mills. After three years of collecting testimony, Mills published a report in December 1907, claiming that baseball had originated in the United States of America and that the first method of playing had been devised by General Abner Doubleday in Cooperstown, New York, in 1839. According to Mills, and to the testimony of Abner Graves, a childhood resident of Cooperstown, Doubleday had altered and thus improved a game similar to rounders called *town ball* by limiting the number of players and by requiring the game to be played by teams. Doubleday supposedly called his new game *baseball.* However, there is no indication in Doubleday's own writing that he had ever been interested in baseball, and in 1839 he was attending the U.S. Military Academy at West Point, far from Cooperstown.

The Doubleday claim was given official confirmation in 1939, when the major leagues made elaborate preparations to commemorate the centennial of the game by establishing the Baseball Hall of Fame in Cooperstown. At that meeting, a legislative committee of the state of New York put into evidence the proposition that Cooperstown was the birthplace of baseball. Furthermore, the Congress of the United States in the 1952 Celler report on baseball continued to maintain that baseball is a game of American origin, and to this day New York State officially claims Cooperstown as the place where Doubleday created modern baseball. The National Baseball Hall of Fame and Museum in Cooperstown itself continues to make much of the so-called Abner Doubleday baseball, an old ball supposedly found in a trunk by a descendant of Graves, asserting that the ball once belonged to Graves and therefore must have been handled by Doubleday as well.

Despite these claims—both official and unofficial—the term *baseball* had been known to both English and American boys long before Doubleday was born. In 1744 a book called *A Little Pretty Pocket Book,* published in London by John Newbery,[2] contained a rhyme describing the game of "Base-Ball" along with a picture illustrating the game. Newbery referred to cricket, town ball, and baseball as distinct games. This book was republished in several American cities in the late 1700s and was reprinted as recently as 1967. Furthermore, according to Harold Seymour,[3] there have been many references to ball games, including baseball, being played in America in the late eighteenth and early nineteenth centuries. George Ewing, a revolutionary war soldier, wrote in his journal of playing the game of "base" on April 7, 1778, at Valley Forge. Thurlow Weed, an upstate New York political boss, wrote in his autobiography that Rochester, New York, had a baseball club that played every afternoon during the ball season of 1825.

To be sure, these games of the eighteenth and early nineteenth centuries were different from the baseball played in the mid-1800s, just as the baseball played in the mid-1800s is different from modern baseball. Children of all ages, Americans and others, are ingenious devisers of sporting games. Many games similar to baseball have been played throughout history. Stickball, played in the streets of many U.S. cities, and the game of "red hot," played by boys and girls in country schoolyards, are games similar to baseball but with the rules somewhat changed to accommodate the number of players, the size of the field, and the equipment available.

Rounders, town ball and baseball appear to have been distinct games at least as early as 1744. The evolution of baseball as we know it probably occurred over many years and over a wide geographical area. Doubleday may or may not have been involved in the evolution, but he certainly did not invent either the game or the name.

The Rules of the Game—Then

Organized baseball as we now know it seems to have come out of men's clubs in the city of New York, specifically the Knickerbockers Baseball Club of New York, organized around 1842 by a group of gentlemen. The Knickerbockers was principally a social club with an exclusive flavor similar to that of country clubs in the early 1900s. To the Knickerbockers, a

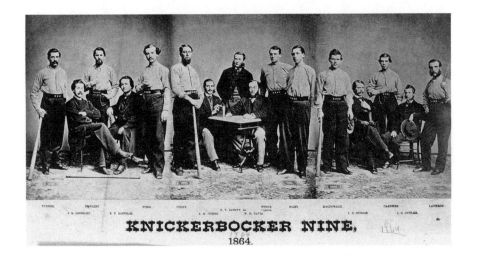

KNICKERBOCKER NINE,
1864.

baseball game was simply a vehicle for amateur recreation and polite social intercourse. The rules of the game emphasized proper and polite conduct. The Knickerbockers apparently had the first written rules of modern baseball. The ball was to weigh 6 to 6.25 ounces and have a circumference of 10 to 10.5 inches (midway in size between a modern baseball and a softball). It had an India rubber center and was wound with yarn and covered with leather, much as is today's ball. A wooden bat of any length, but not over 2.5 inches in diameter at its thickest part, was used. The players took their positions on a square field with 90-foot baselines, three bases and a home plate. The pitcher stood 45 feet from the home base. For the first time, batters had to take their turn in a regular rotation. Innings were over after three outs were made, but games could last any number of innings. The first team to score 21 runs was the winner. A batter was declared out if he swung at and missed three pitches, provided the catcher caught the last one either on the fly or on the first bounce. He was also out if a fair ball was caught on the fly or a foul ball was caught on the fly or on the first bounce. An important change was the elimination of the old practice of "soaking." In the games of rounders and town ball a batter was put out if he was hit, or "soaked," by a ball thrown at him when he was off base. In the Knickerbocker rules, this was replaced with the new rule of putting out the runner by touching him with the ball.

The pitching rules were closer to those of modern slow-pitch softball than to those of modern baseball. The pitcher had to toss the ball gently underhand as near to the plate as possible. The batter could ask for a high pitch or a low pitch and decline to swing at as many pitches as he wished. Integral were the postgame festivities, frequently including lavish dinner dances.

The All-American Sport

The early baseball clubs were restricted to upper social classes. Baseball itself could hardly have remained so restricted, however, because the game was popular and inexpensive, and in the cities of the time, there were many public parks and vacant lots. Part of baseball's present-day popularity lies, no doubt, in the timeliness of its origin. According to John R. Betts,[4] the rise of the sport in America during the era of the Industrial Revolution is not coincidental. The most popular American sports of the early 1800s—horse racing, fishing and fox hunting—were distinctly rural. By the middle of the century, many of the white-collar workers who lived in the rapidly expanding cities had become concerned about their sedentary lives. Hence, the emergence of baseball as an activity of the upper social classes. Easy access to inexpensive newspapers and the development of rail transportation led to rapidly increased interest in organized sports, by both participants and spectators. But why did baseball, as opposed to some other organized sport, emerge as America's national pastime? Horse racing was a popular spectator sport, but it did not allow mass participation. Basketball was not invented until 1892. Football has always been stigmatized by excessive brutality. As we have already said, baseball was inexpensive to play, and the availability of vacant lots and public parks made it easy for people to participate in the game. The rules were relatively easy to learn, and, in any case, most adults were familiar with them because they had played some version of the game in their childhood.

Above all, baseball is a sport where the fans can identify with the players, who appear to be men of average size and strength. To be sure, playing the game at its highest level requires tremendous skill, but in ways that can be emulated by most of us—at least in modest ways. Few have the hand-eye coordination of Wade Boggs or Tony Gwynn, but most of us can hit or catch a ball thrown or batted by one of our peers at the annual Labor Day picnic. Football is dominated by men of extraordinary size and strength. Few can truly identify with the exploits of Greg Lloyd, Bruce Smith, Simeon Rice or Levon Kirkland. These men are fundamentally different from the rest of us. Similarly, basketball is dominated by extraordinary tall men. In the National Basketball Association (NBA), Larry Bird displayed considerable skill, but would this suffice if he were not a foot taller than the average American male? Every NBA team is searching for the franchise player: the one person that will produce financial success for the organization. In basketball, put a good team around Kareem Abdul Jabbar or Michael Jordan and you get a championship. Michael Jordan is simply an extraordinary athlete. His presence changes a team completely, as his absence from the Chicago Bulls testifies. In baseball no single person normally dominates. It is a true team sport. In baseball each player must play his part, each must take his turn at hitting and at fielding. Nobody participates in every play, except the pitcher and the catcher. Unless the batter strikes out, others must cooperate to get an out. Thus, cooperation and team play are of great importance. But most important, the game is played mostly by men who at least appear to be physically normal. And the presence of Mark McGwire and Sammy Sosa, men of extraordinary strength, does not guarantee their teams a championship.

Professional baseball players are not, of course, ordinary men. Consider, for example, the task that must be performed by a batter playing baseball in the major leagues. His task is to hit a 2.85-inch diameter ball with a bat whose thicker end is 2.75 inches in diameter. The ball is thrown by a pitcher whose sole purpose is to prevent the batter from hitting the ball solidly. He throws the ball from a distance of 60 feet, 6 inches from the batter and at speeds ranging from perhaps 70 to 100 miles per hour. More often than not, the ball is spinning at 2000 revolutions per minute and might be curving in one direction or another. The batter has perhaps half a second to decide to swing and to direct his bat to make a solid collision with the ball. Even if he hits it solidly, if the direction of the ball is such that an opposing player can either catch it before it reaches the ground or can relay the ball to first base before the batter runs there, the batter has failed in his task. If the batter succeeds just once in three tries, he is a star. Think of another sport in which that is true.

The Rules Change

Although the game the New York Knickerbockers played in the 1840s would be recognizable to today's baseball fan, both the rules of the game and the way it is played have undergone significant changes over the decades. In the early game, basemen actually played on the base, and the pitcher worked from a 6-foot-square box only 45 feet from home plate. The ball had to be thrown underhanded in such a manner that the hand always passed below the hip. The catcher stood some 20 feet behind the batsman without any of the protective devices used today. Indeed, no gloves or other protective devices were used by any of the players before about 1875. In that year, the first glove, a skintight glove with the finger-tips cut off, was reputedly worn by Charles C. Waitt, a first baseman for the St. Louis team. Catching was a tougher job than it is today, especially after the catcher moved up closer to the batter to be in a better position to throw out runners trying to steal. In the 1870s, the catcher's mask was introduced by Fred Thayer, the captain of the Harvard team. Before 1864, balls caught on the first bounce put the runner out. In that year, the fly rule was introduced, requiring that for an out, batted balls must be caught on the fly. As the game became more than a polite exercise in sportsmanship, and winning became more important, players began to devise tricks to circumvent the rules. Dickie Pierce, a shortstop for the Atlantics, introduced bunting by gently tapping the ball in front of home plate and running to first base before the bewildered opponents could recover. For years this was considered unsportsmanlike, eliciting boos and catcalls from the fans. Fielders began dropping fly balls intentionally to convert an easy out into a double play. This necessitated the adoption of the infield fly rule. Basemen moved off their bases to better defensive positions. The art of backing up other fielders took hold. Sliding to avoid being tagged became common.

The rule-makers recognized early on, however, that the game was essentially a duel between the batter and the pitcher, and many of the rule changes reflected this. Pitchers began accumulating tricks to outwit the batters. Jim Creighton of the Brooklyn Stars and others began cheating on pitching style, which required the ball to be delivered with a stiff-armed, underhanded motion. Creighton began speed pitching by adding an almost imperceptible wrist snap to his delivery. Creighton and Harry Wright initiated the change-up, a slow pitch thrown like a fastball. In the 1860s or 1870s the curve ball was introduced. The inventor of the curve will probably never be known for certain, although Arthur "Candy" Cummings may have been its originator. Cummings, who pitched for the Brooklyn Stars in the 1860s, is traditionally credited as the first pitcher to throw a curve ball, and a bust of him is in the Baseball Hall of Fame. However, Fred Goldsmith, an early pitcher, claimed that Charles Avery, a

Yale player, invented the curve. Later, Goldsmith claimed he himself had invented the pitch. According to Robert Smith,[5] many pitchers in the 1860s were using the curve ball, but Cummings may have been the first to truly cultivate the pitch. Allison Danzig and Joe Reichler[6] state that Creighton was throwing the curve ball before 1862. Indeed, Henry Chadwick claimed that a Rochester pitcher, whose name he could not remember, was using the pitch as early as the 1850s. The curve ball may have had several independent inventors.

In the 1870s, as underhanded pitching improved, fans began to complain of dull, low-scoring games. Rule changes favoring the batter were instituted to make the game more interesting to the fans. In 1879, the size of the pitcher's box was reduced to 6 by 4 feet, and the pitcher, as he started his motion, was required to face the batter. The number of balls required for a base on balls was reduced from 9 to 8. The batter would be called out on strikes if the catcher caught the third strike on the fly, which required the catcher to move closer to the plate. In 1881, the pitching distance was lengthened from 45 to 50 feet. In 1884, pitchers were allowed to throw with a shoulder-high motion, ushering in the era of overhand pitching. In 1885, the batters were aided when a rule was initiated allowing the use of a flat-sided bat. In 1886, the pitching box was enlarged to 7 by 4 feet. In 1887, batters were no longer given the privilege of requesting a high ball or a low ball. To compensate, four strikes were allowed if the third strike was a called strike, and a walk was counted as a hit. Batters hit by a pitched ball were given first base, and pitchers were forbidden to lift a foot until they actually delivered the ball. In 1893, the pitching distance was lengthened to the present distance of 60.5 feet. In 1893, the pitcher's rubber was introduced and the round bat was again required.

Blacks in Baseball [7,8]

In 1947, did Jackie Robinson become the first African American in organized baseball? Not quite. He was the first black person to play in organized baseball in more than fifty years, but he was not the first ever. (By "organized baseball" we mean the official major leagues and their minor league affiliates.)

While some evidence exists for black teams playing baseball games in the early nineteenth century, the first records of black teams emerged in 1867 when the Uniques of Brooklyn played the Excelsiors of Philadelphia, with the Excelsiors emerging victorious. That same year the Pythians, another black Philadelphia team, requested membership in the newly organized National Association of Baseball Players. The request was unanimously rejected because, "If colored clubs were admitted there would be, in all probability, some division of feeling, whereas, by excluding them no injury could result to anyone." In 1876, when the National League was established, there was no written agreement to exclude blacks, but they were nevertheless systematically excluded through a "gentleman's agreement."

But as many as a few dozen blacks managed to play in the minor leagues, and even sporadically in the major leagues, although not without a good deal of animosity from white players, managers and fans. Perhaps the first to break into the minor leagues was Bud Fowler, who played both pitcher and second base for a number of independent and semi-professional teams, including the Lynn Live Oaks of the International Association, the first minor league. The magazine *Sporting Life* reported that "If he had a white face he would be playing with the best of them. . . . Those who know, say that there is no better second baseman in the country."

In 1887, another great second baseman, Frank Grant, joined the Buffalo team in the International Association. The light-skinned Grant, described as a Spaniard by the *Buffalo Express*, batted .350, third best in the league.

Moses Fleetwood Walker, son of an Ohio physician and a student at Oberlin College, was a barehanded catcher who played for both Oberlin and the University of Michigan in the early 1880s. In 1883, he appeared in 60 games for Toledo in the Northwestern League. The following year Toledo joined the American Association, making him the first black to play in the major leagues. He was joined there that same year by his younger brother, Welday. The following year Toledo dropped out of the American Association, ending the major league careers of the two black brothers.

Confrontations, insults and threats from white players, fans and managers were legendary. In Louisville, fans hissed and screamed racial slurs at Walker, and in Richmond, he received death threats. Manager Cap Anson of the Chicago White Stockings declared that he would not play an exhibition game against Toledo if Walker played. (Anson was one of the greatest hitters of the nineteenth century, batting over .300 in 24 of his 27 seasons as a player, but he was a bigot with a bad temper.) Toledo manager Charlie Morton inserted Walker into the lineup anyway and the game was played.

By 1887, Walker, Fowler and a black pitcher named George Stovey had joined the minor league Newark team in the International League. Black players were increasing in numbers in that league, with six of the 10 teams fielding blacks. Stovey set an International League record that still stands, winning 35 games that year. Frank Grant, second baseman for the Buffalo team, won the batting title and led the league in home runs, while Fowler hit .350 and stole 23 bases for the Binghamton team.

Fans continued to shout insults and threats at the black players. Local newspapers often displayed headlines declaring that playing with black players was distasteful. Pitchers would constantly throw at black hitters. When a black pitcher took the mound, his white teammates would often purposely field badly. Eventually, Fowler's teammates refused to take the field with him, and soon after the Binghamton team removed him and another black pitcher.

The Syracuse team in 1887 included a light-skinned catcher who attempted to pass himself off as a white man named Dick Male. His real name was Richard Johnson. He was released after a poor early performance and joined the Zanesville club in the Ohio State League under his real name.

In July, the Newark team planned to send Stovey to the mound against the Chicago White Stockings in an exhibition game. Again, Anson refused to play an exhibition game if Stovey played. This time Stovey was benched. By the end of the 1887–1888 season, every black player in the top five minor leagues had been told that his contract would not be renewed. Although no formal rule was adopted, white players and managers had apparently decided to segregate the game. Although some lower minor leagues continued to employ a few black players, the door had effectively been slammed shut for blacks on integrated teams in organized baseball.

All-black teams were allowed to compete with all-white teams for a decade, but by the 1890s blacks were relegated to all-black teams, playing most of their games on barnstorming tours, outside the realm of organized baseball.

There were a few attempts to reintroduce black players into major league teams by claiming that they were not blacks. In 1901, John "Muggsy" McGraw, then manager of the Baltimore Orioles, tried to sign a second baseman named Charlie Grant by claiming that he was a Cherokee Indian named Charlie Tokohoma, but everyone saw through the trick and Grant was never signed. McGraw tried again in 1906, this time as manager of the New York Giants. He talked of signing Cuban pitcher Jose Mendez and the great black pitcher Rube Foster but was talked out of it by white owners. In 1911, two light-skinned Cuban players, third baseman Rafael Almeida and outfielder Armando Marsans, were actually signed by the Cincinnati Reds. It was soon clear, however, that darker-skinned Cubans like Mendez would be passed over. In the fall of 1908, the Reds toured Cuba and played a series of games against all-black teams. In the first game, Mendez threw a one-hitter, shutting out the Reds 1−0. He shut them out again in two subsequent games, pitching 25 consecutive scoreless innings against one of the best major league teams of that era.

Again in 1919, Rube Foster put together a plan to integrate organized baseball, proposing to baseball commissioner Kenesaw Mountain Landis that one all-black team be admitted to each of the two major leagues. He was turned down. The following year, black owners converged on Kansas City and formed the Negro National League. Other black leagues followed.

It can perhaps be said that the Negro leagues flourished for awhile, but there were mitigating factors that prevented eventual financial success. The teams depended on playing in major or minor league parks, which they rented for their games when the home teams were out of town. As a result, scheduling appears to have taken place sporadically, and advanced publicity was nearly impossible. Talent was uneven. Written contracts did not exist, and the best players jumped from team to team, sometimes playing for two or three different teams in one year, going where the money was best at a moment's notice.

There is little doubt that some of the best players in the history of the game played in the Negro leagues before Jackie Robinson broke the color barrier. Many of the names are now familiar to baseball fans, like Hall of Famers Satchel Paige, Josh Gibson, Judy Johnson, Ray Dandridge, Cool Papa Bell, Monte Irvin and Pop Lloyd, but there were many, many others with major league talent. All-star teams from the Negro leagues frequently played traveling groups of major league teams such as the Babe Ruth All-Stars and the Dizzy Dean All-Stars, and from all appearances held their own or proved superior. By some accounts, the black all-stars won as many as two thirds of the games. Much has been written about the great black players of that era, but most of it cannot be documented. Most major cities had at least one newspaper devoted to the black community, but coverage of box scores of baseball games was inconsistent. Unfortunately, we are left mostly with anecdotal information, as with tales of Buck Leonard hitting a home run off Bob Feller that cleared the bleachers and a row of houses and hit a water tower, or how Josh Gibson hit a home run into the upper deck of Yankee Stadium with one hand. Some are clearly intended to be humorous, such as the tale of Judy Johnson being so fast that he hit a hard grounder through the pitcher's mound and it hit him in the behind as he reached second base. Nevertheless, there is no doubt that whatever their exploits, many of these players had major league talent.

In 1943, Bill Veeck attempted to buy the Philadelphia Phillies, planning to raid the Negro leagues and stock his new team. Rumors circulated, and when they reached Commissioner Landis the team was suddenly sold to the National League, which then sold the team to William Cox, at a substantial loss. Landis publicly insisted that there were no players in the Negro leagues with the talent to play major league baseball. The absurdity of the argument was evident; during World War II, while no black player made it to the major leagues, a one-armed white player, Pete Gray played outfield for the St. Louis team in the American League.

Despite Landis and those who shared his prejudices, the mood in the United States began to shift by the 1940s. But the owners continued to worry about what effect allowing blacks to play on their teams would have on fan support and finances. Not a few of the players themselves were outspoken racists. But Landis died in 1944 and A. B. "Happy" Chandler became the baseball commissioner, and he was an outspoken advocate of allowing blacks into major league baseball. Another advocate, Branch Rickey, had become president of the Brooklyn Dodgers, and informed the Dodger owners that he intended to recruit black players in the near future. Still, when the Brooklyn Dodgers added Jackie Robinson to their team, they moved their spring training camp to Havana for the 1946 and 1947 seasons. Robinson had been signed to a three-year contract by Rickey and assigned for his first year to play second base for the Montreal minor league team. There he led second basemen in fielding percentage and runs scored, batted a league-leading .349 and finished second in stolen bases. The following year, playing for the Dodgers, he led the National League in stolen bases while batting .297 and receiving Rookie of the Year honors.

Robinson's success paved the way for more stars from the Negro leagues to join major league teams, but the going was slow. Many of the great players, like Buck Leonard, Josh Gibson and Cool Papa Bell, were considered too old. Satchel Paige was eventually signed by Cleveland, probably as a publicity stunt, when he was reportedly 42 years old (nobody ever knew his actual age). Paige responded by pitching so well in relief that many thought he should have received the Rookie of the Year award. By 1948, the Dodgers had signed Roy Campanella, Don Newcombe and Joe Black, and the Cleveland Indians had signed Larry Doby, Luke Easter and Paige. Both Newcombe and Black earned Rookie of the Year honors. By the mid-1950s, the major leagues had taken most of the younger talent from the Negro leagues and had begun to sign players like Bob Gibson and Frank Robinson who had never played in those leagues. The Negro leagues were doomed. They continued to play with second-rate talent for awhile, but by 1965, only one team remained: the Indianapolis Clowns, a team that was mostly for comic relief.

Women in Baseball[9]

Many associate the beginning of women in baseball with the movie *A League of Their Own* (1992). In 1943, with the beginning of U.S. participation in World War II, the shortage of male baseball players threatened major league baseball. The Chicago Cubs owner formed a women's softball club to play in major league stadiums while male players were away at war. The league soon modified its rules, and the All-American Girls Baseball League (AAGBL) was formed. In the beginning there were four teams, the Rockford Peaches, the South Bend Blue Sox, the Racine Belles, and the Kenosha Comets. More teams were added until the number reached 10 in 1948, and in that year the league's games drew in a million fans. During its 12-year existence, more than 500 exceptional athletes played in the AAGBL, and they were coached by some of the giants of major league baseball, men like Bill Wambsganss, Jimmy Fox, Max Carey and Dave Bancroft. Once the war ended, social pressure grew for women to return to domestic duties, and the AAGBL folded.

Although this may have been the high point of women's active participation in baseball, it is far from the beginning or, for that matter, the end of that participation. Owners had hosted regular "Ladies Days" to add to the gate receipts, but some women longed to participate as players. Those who did so had to endure derision similar to that aimed at black players. In September 1875, a group of men organized a women's baseball club in Springfield, Illinois. Two teams, called the Blondes and the Brunettes, played a match on September 11 of that year. Another group in Philadelphia organized the Young Ladies' Baseball Club in 1883. The games were generally considered entertainment spectacles, not as serious competition, but as comic relief. The management claimed they had selected girls from 200 applicants, but when the first game was played on August 18, 1883, there were only 16 girls on the two teams, and the roster had to be filled out by recruiting two men.

In the late nineteenth and early twentieth centuries, several women's teams traveled from town to town playing semiprofessional and amateur men's teams. Among the best known were the Bloomer Girls teams. Many of these teams were made up of people of both sexes. Male members included Smoky Joe Woods and Rogers Hornsby, future major league stars. There were also teams at women's colleges. Vassar College had two organized teams as early as 1866. In about 1907, Alta Weiss reportedly financed her college education by pitching for the all-male Vermillion Independents, a semiprofessional team in Ohio. In 1928, 14-year-old Margaret Gisolo helped lead a Blanford, Indiana, American Legion team to the state championship. Opposing teams protested to the American League Commission, which referred it to major league Commissioner Landis, who disallowed the protest. In 1931, the Chattanooga Lookouts of the minor league Southern Association signed a 17-year-old female pitcher named Jackie Mitchell, making her the first female professional player. After pitching in an exhibition game against the New York Yankees and striking out Babe Ruth and Lou Gehrig in succession, she became an overnight celebrity. This time Landis was more decisive. He disallowed Mitchell's contract, stating that life in baseball was too strenuous for women. In 1952, Commissioner Ford Frick formalized the rule banning women from the professional game, ostensibly to prevent teams from using women as publicity stunts. There seems to be less concern for girls playing on Little League teams, with 1998 marking the twenty-fifth season since this practice began.

In 1994, Coors Brewing Company sponsored the Colorado Silver Bullets, which became the first women's baseball team officially recognized by the National Association of Professional Baseball Leagues. The team struggled for awhile, competing with all-male teams, but by 1997, it had its first winning season. However, the team folded that year when Coors did not renew its sponsorship. Ladies League Baseball began with four teams, one in San Jose, one in Phoenix and two in Los Angeles, in 1997 and expanded to New Jersey and Buffalo the following year. Low attendance and financial problems forced its cancellation in the middle of the 1998 season.

Interest by women in playing baseball continues to grow, and it appears only to be a matter of time before some success is achieved in organizing a professional league and, yes, of women appearing on all-male teams in organized baseball.

A Tricky Pitch

Following the rule changes in the 1880s and 1890s, pitchers began to experiment with making the ball do tricks. According to Danzig and Reichler, Billy Rhines of the Cincinnati Reds was the first to throw a rise ball, and Tom Ramsey was the originator of the sinker. Christy Matthewson later began to throw the screwball, a reverse curve that he called the "hideaway."

Elmer Stricklett in 1902 was the first to throw the spitball, a pitch he had apparently learned from George Hildebrand, an outfielder who was his teammate on the Sacramento Club. Stricklett himself later claimed that Frank Corridon of Providence began throwing the spitball before he did.

Eddie Cicotte was the first pitcher to use the knuckleball, according to Danzig and Reichler. However, as with most other pitches, the real originator of the knuckle ball seems somewhat obscure. Martin Quigley states, "Pitchers tried their hand at the knuckle ball to replace the spitball that had been outlawed in 1919. But the origin of the knuckler goes way back. Ed Summers of the Detroit Tigers, who used it unsuccessfully in the games against the Cubs in the World Series of 1908, said he learned it from an old timer who threw it sidearm in the early '80s."[10]

Possibly the earliest consistent practitioner of the knuckle ball in the major leagues was Ed Rommel of the Philadelphia Athletics. Rommel used the pitch to win 27 games for the seventh-place Athletics in 1922. Even after the spitball was outlawed in 1919, it continued to be used surreptitiously by major league pitchers. In his book *The Crooked Pitch* (1984), Quigley provides a fine description of the spitball as well as of a host of other illegal pitches, including the mud ball, the shine ball and the emery ball. He also provides a quote from Ray Miller, then the pitching coach of the Baltimore Orioles and one of the most colorful men in baseball: "Someday, I expect to see a pitcher walk out to the mound with a utility belt on—file, chisel, screwdriver, glue. . . . He'll throw the ball to the plate with bolts attached to it."

A favorite trick of the old-timers was to secretly use frozen baseballs. Steven Hall[11] claims that "this is exactly what the good-pitch, no-hit

Chicago White Sox used to do when hard-hitting teams came to town in the late 1950s." The coefficient of restitution, that is, the bounciness, of the India rubber center of a baseball decreases rapidly as the temperature goes down. The frozen ball simply could not be hit as far.

A Tricky Bat

Hitters perform their mischief, too. The most common illegal act performed by the batter is the "doctoring" of the tool of his trade: the bat. A common trick is drilling a hole in the thicker end of the bat along its length and filling it with something light, typically cork. This makes the bat lighter and therefore easier to swing, producing faster bat speed and allowing better control. In Chapter 5, we discuss the effects of corking the bat. Former Yankee shortstop Tony Kubek claims that the players of his day used an ice pick to carve out the bat's dark grain, which is the softer part of wood. They would then put pine tar in the grooves, let it harden and resand it. This was supposed to give the bat a harder surface. There are also stories of players leaving the grooves in the bat after picking out the dark grain, thereby roughening the surface of the bat. The batter is less likely to foul the ball back or to hit a pop-up if the surface of the ball does not slip too much on the surface of the bat. Tampering with the bat in ways such as this is no longer officially allowed, but this does not mean that such things are no longer done.

Wet pine tar is a sticky substance that batters put on their bats to improve their grip. Placing pine tar beyond 18 inches from the small end of the bat is not allowed. In 1984 there was a controversy involving George Brett of the Kansas City Royals in a game against the New York Yankees. After Brett hit a home run, Yankee manager Billy Martin brought the bat to the umpire's attention, demonstrating that the pine tar had gone beyond the 18-inch legal limit. The homer was nullified. American League President Lee McPhail later reinstated the homer, saying the rule was intended simply to keep the ball clean. George Brett probably knew better.

Baseball and Science

Baseball players are constant practitioners of applied physics. Each time a pitcher throws a ball or a batter alters the surface of his bat, he is putting the laws of physics to work for his benefit. Once you have read the remainder of this book, you will have a better understanding of how this is done.

Notes

1. William Clarke, *The Boy's and Girl's Book of Sports*, J. S. Hammond, Providence, R.I., 1841. An early book about sporting games for children.

2. John Newbery, *A Little Pretty Pocket Book*, Harcourt Brace and World, New York, 1967. A book of sports games first published in the eighteenth century.

3. Harold Seymour, *Baseball*, Oxford University Press, New York, 1960.

4. John R. Betts, *America's Sporting Heritage, 1850–1950*, Addison-Wesley, Reading, Mass., 1974. A scholarly history of sports in the United States.

5. Robert Smith, *Baseball*, Simon and Schuster, New York, 1947.

6. Allison Danzig and Joe Reichler, *The History of Baseball: Its Great Players, Teams and Managers*, Prentice-Hall, Englewood Cliffs, N.J., 1959. Contains interesting historical notes, good photographs and anecdotal stories about great teams and individual players.

7. Jules Tygiel, "Blackball," in *Total Baseball*, edited by John Thorn et al., Total Sports, New York, 1999.

8. Bruce Chadwick, *When the Game Was Black and White: The Illustrated History of Baseball's Negro Leagues*, Abbeville Press, New York, 1992. A fascinating history of the Negro leagues in words and pictures.

9. Debra Shattuck, "Women in Baseball," in *Total Baseball*, edited by John Thorn et al., Total Sports, New York, 1999.

10. Martin Quigley, *The Crooked Pitch: The Curve Ball in American Baseball History*, Algonquin Books, Chapel Hill, N.C., 1984.

11. Steven S. Hall, "Baseball's Dirty Tricks," *Science 83*, vol. 4, no. 8, August 1983, pp. 92–93.

    Paul Dickson, *The New Dickson Baseball Dictionary*, Harcourt Brace and Company, New York, 1999.

The Flight of the Ball:

Do Curve Balls Curve?

"A curve ball," says Baltimore Orioles pinch hitter Terry Crowley, "comes straight in, then about 4 or 5 feet from home plate, it breaks straight down." "It doesn't look like the ball starts breaking until about 10 feet away," agrees Oriole catcher Joe Nolan. According to W. F. Allman,[1] Rich Dauer of the Orioles does not believe it: "A curve is a big swoop. It is like you have a rock on a string and you swing it. It goes out, and it curves in. Like you drew it with a compass. It's always the same rainbow."

Sandy Koufax's curve ball has been described as looking as if it rolled off a table. Legend has it that Walter Johnson could make his fastball rise, appearing to defy the law of gravity. The flights of curve balls and "rising" fastballs, as well as many other pitches—legal and illegal—are the subjects of much baseball folklore. Does a baseball curve? If so, does it fly in the "rainbow" arc described by Dauer, or does it "break," as many hitters believe? Is a "rising" fastball possible, even when the man throwing it is the great Walter Johnson?

It may be especially surprising to those of you who have hit (or swung at) a well-thrown curve ball that there has not always been universal agreement on whether a ball curves at all, let alone whether it "breaks" or curves in a rainbow arc. According to All-man, "In an 1870 demonstration Fred Goldsmith reportedly threw a sidearm curve at three upright poles arranged in a straight line. The ball traveled to the right of the first, to the left of the second, and to the right of the third," thus proving that a curve ball curves. Will White of the old Cincinnati Red Stockings, "one of the league's best right-handers," reportedly accomplished this same feat in 1877, as Charles Einstein[2] tells it. Still doubts persisted. A Professor Stoddard of Worcester University in Massachusetts wrote in the *Cincinnati Inquirer,* shortly after White's demonstration, that "it is not only theoretically but practically impossible for any such impetus to be conveyed to a moving body as would be required . . . to control the movement of what is termed a curve ball."

In 1941 the controversy began anew, when an old baseball scout claimed that there was no such thing as a curve ball, and that everyone in baseball knew it (see Einstein[2]). This story was contained in a letter to *The New Yorker* magazine from R. W. Madden,[3] in which he recounted a conversation with an elderly gentleman who claimed to be a major league scout and asked not to be named because, "The guys in baseball have an interest in keeping the secret. It makes the game exciting to have curves. In the second place, some guys might be willing to say they doubted if there was such a thing as curves, only what would happen to 'em? I'll tell you. They'd be laughed out of town." The response from readers lasted for several months. Noel Houston[4] wrote that he recalled that Dick Rover, a fictional character in the famous Rover Boys books, had "proved" to an elderly professor that curve balls do curve by throwing a curve ball through a series of wooden frames covered with dampened tissue paper. When the frames were later laid on top of one another, they showed clearly that the ball had traveled in a straight line until it neared home plate, after which the sudden veering of the positions of the holes traced the break of the ball. Lucy Stanton[5] wrote that her father had seen a demonstration by Tommy Bond, a pitcher with the Boston Baseball Club. Bond threw a curve ball that weaved between three stakes just as Fred Goldsmith and Will White were supposed to have done; this demonstration was also reported by Robert Smith.[6] However, Ray D. Lillibridge disputed all these experiments in 1941, reporting that one man, after attending such a demonstration, suggested that "we were hypnotized . . . [because] the human eye could not detect which side of a stake an object the size of a base-ball traveling at that speed had passed."[7]

A Picture's Worth a Thousand Words

In the same year that Lillibridge disputed the feasibility of detecting curve balls, *Life* magazine picked up the gauntlet. The magazine's high-speed photographer, Gjon Mili, using high-speed stroboscopic lighting at night, positioned three cameras to photograph balls thrown by two pitchers: Cy Blanton of the Philadelphia Phillies and Carl Hubbell of the New York Giants. The conclusions drawn from the resulting photographs were that Blanton's "best curve is really a straight ball with a decided drop," and that "Hub threw everything in his pitching book" and "wound up with . . . two straight lines." Although the publishers failed at the time to recognize the fact, the pictures published in *Life* magazine are the first hard proof of the existence of the curve ball. The pictures taken by Mili are shown in Figure 1. Once we have set the stage with a little background science, we will return our attention to these pictures and discover why they prove the existence of the curve ball.

The Laws of Motion

Before we begin our analysis of the curve ball, let us first look at a simpler situation. Our everyday experience tells us that objects such as the ball in Figure 2 will not move unless pushed or pulled, that is, unless acted on by a force. We call this tendency of objects at rest to remain at rest *inertia*. One measure of inertia is mass. The greater the mass of a body, the greater will be its tendency to remain at rest.

Now let us analyze what happens to the ball when it is moving at a constant speed toward the other end of the table. If we neglect for the moment the effects of friction between the ball and the table, the ball will maintain its *uniform motion* as it rolls across the table. As the length of the arrows in Figure 3 indicate, it will tend to maintain the same speed at which it started its journey.

These observations can be summarized as follows:

Unless pushed or pulled, an object or body in a state of rest will tend to remain at rest. Likewise, a body in motion at a particular speed and in a particular direction will tend to maintain the same speed and travel in the same direction.

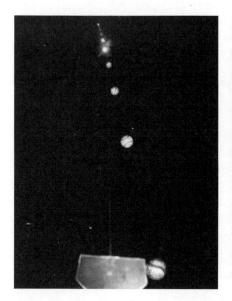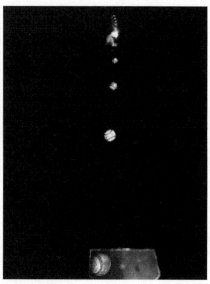

Figure 1. The first attempt to use photography to prove that a curve ball really curves was done by Life magazine in 1941. The balls shown here were photographed at night using stroboscopic lighting, giving a series of stop-action photos of the baseball. At the left is a pitch considered a curve ball and at the right is a pitch considered a screwball, both thrown by Carl Hubbell of the New York Giants. The editors of Life concluded that neither ball curved, because both trajectories appear to lie in a plane. (Photographs, Gjon Mili, Life magazine © 1941 Time, Inc.)

Now let us see what happens when we push the moving ball from another direction. As Figure 4 shows, both its speed and direction of motion will change. With this observation, our previous statement can be qualified:

A body at rest or in a state of uniform motion in a straight line will remain at rest or maintain its speed and direction of motion unless acted on by a force.

Although our observations of the ball's motion may not seem to be very important, they constitute one of the most fundamental laws in physics. In fact, it is called the *first law of motion*. It was first formulated by the seventeenth-century physicist Isaac Newton, who was the first to recognize

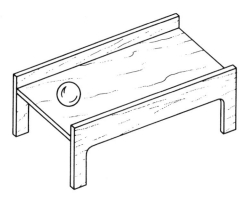

Figure 2.   A ball with a smooth surface sitting on a level table will not move unless acted on by a force.

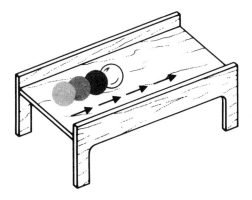

Figure 3.   The ball rolling across the table will maintain its uniform motion.

its fundamental applicability to almost every aspect of the physical universe. To honor this scientific genius, it is also called *Newton's first law of motion.*

We have been talking about the speed and direction of a moving body. In physics, however, the term *velocity* is used to describe these two aspects of motion. Thus, we speak of a velocity of 80 miles per hour *due east.* One way to graphically show the velocity of a moving object is to

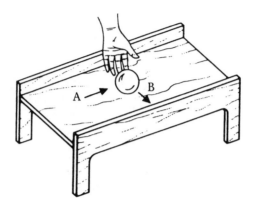

Figure 4. *When the rolling ball is pushed from another direction, both its speed and direction of motion are changed.*

use a vector. (See boxes, pages 26 and 27.) In Figure 5, the arrows are vectors representing the velocity of the ball at points *A* and *B*. The lengths of the arrows indicate the magnitudes of the velocities, that is, the speed of the ball at points *A* and *B*, while the directions in which the arrows point indicate the directions of motion of the ball at each point.

Another important vector quantity is *acceleration*. As we know very well from our personal experiences on public transportation or on the highway, acceleration is a rate of change in the velocity of a moving body. Thus, if a train traveling south at 40 miles per hour changes its speed to 60 miles an hour in 2 seconds, it experiences an average acceleration of 20 miles per hour per 2 seconds, or 10 mi/h/s in the direction of south. The acceleration is the rate at which the velocity changes with time. Thus, the average acceleration experienced by the ball in Figure 5 is the change in the velocity vector divided by the time required to go from point *A* to point *B*. (See box, page 29.)

Keep in mind that a change in acceleration does not always require a change in the speed of an object. For example, although the speed of the ball at points *A* and *B* in Figure 6 is the same, the direction in which the ball is traveling is different at each point. Thus, the velocity of the ball has changed in going from point *A* to point *B*—that is, the ball has accelerated.

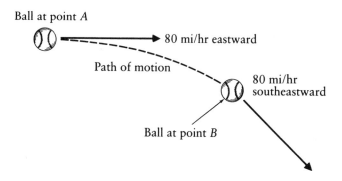

Figure 5.   Although the ball is moving at the same speed at points A and B, its velocity has changed because velocity is a vector. A vector has both magnitude and direction, and the direction of the velocity vector changed as the ball traveled from point A to point B.

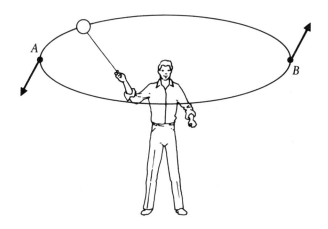

Figure 6.   Although the speed of the ball is the same at points A and B, the directions of motion and thus the velocities are different.

One more important vector quantity is force. When we want to specify the force experienced by an object, we must use a vector to indicate not only the size of the force but also the direction in which it acts. Since weight is a force experienced by all material bodies on Earth, it must be graphically represented by a vector, a vector pointing toward

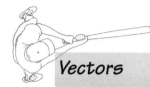

Vectors

quantity that can be specified by a magnitude only is called a *scalar*. If a particle undergoes a change in its position, its new position and old position are related by two quantities, the distance between the two quantities, and the direction of the second position relative to the first. The change in position is called *displacement*. It has both magnitude and direction. It can be represented by drawing an arrow from the first position (position *A*) to the second position (position *B*). The length of the arrow represents the distance, or *magnitude*, of the displacement.

If the particle subsequently moves to another position (position *C*), the displacement from *B* to *C* can be constructed by drawing an arrow from *B* to *C*. Furthermore, the total displacement from *A* to *C* can be represented by an arrow drawn from *A* to *C*. The sum of the first two arrows is represented by the third arrow. Quantities that behave like this are called *vectors*.

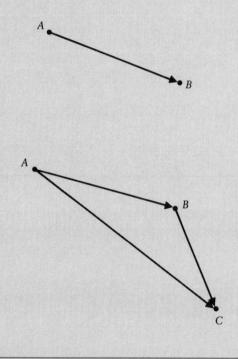

Adding and Subtracting Vectors

et us see how the displacement vector between points B and C can be added to give the displacement vector between points A and C. First we name the displacement from A and B **D**(AB) and the displacement from B and C **D**(BC). We use boldface type to represent the vectors **D**(AB) and **D**(BC). The directions of the two vectors are specified by the two angles θ(AB) and θ(BC) in the figure. Because of the way vectors are added, we can represent **D**(AB) by the sum of the two vectors with magnitudes **D**(AB) cos θ(AB) pointing in the horizontal direction and **D**(AB) sin θ(AB) pointing downward. These are called *components* of the vector. Similarly, the vector **D**(BC) can be represented by **D**(BC) cos θ(BC) pointing to the right and **D**(BC) sin θ(BC) pointing downward. Now colinear vectors (vectors pointed in the same or opposite directions) can be added directly. Therefore, the vector

$$\mathbf{D}(AC) = \mathbf{D}(AB) + \mathbf{D}(BC)$$

has a horizontal component **D**(AB) cos θ(AB) + **D**(BC) cos θ(BC) and a vertically downward component of magnitude **D**(AB) sin θ(AB) + **D**(BC) sin θ(BC).

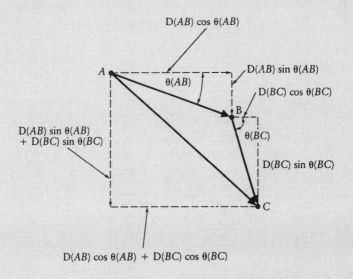

Earth. Now that we are armed with this background, let us return to the ball rolling on the table.

As Figure 7 shows, there are two forces acting on the rolling ball. One is the force of gravity. This force pulls the ball downward toward the ground. We call this the *weight* of the ball, and give it the symbol *w*. Since the ball does not move toward the ground, there must also be a force equal in strength pushing up on the ball. This force is exerted by the table surface on the ball. As the figure shows, these two forces are equal in strength but opposite in direction and thus cancel one another. That is, there is no net force on the ball. According to Newton's first law, therefore, it will continue to roll at the same speed and in the same direction.

Suppose the ball rolls off the surface. The upward force exerted by the surface on the ball will not be present, and our experience tells us that the ball will be accelerated downward, or fall, because it experiences an unbalanced force, the weight of the ball. To understand the resulting motion, we again turn to Newton for another law of motion:

The acceleration of a body that is acted on by a net force is inversely proportional to its mass and directly proportional to the net force acting on it.

This is called the *second law of motion* or *Newton's second law*. (See box, page 30.)

Now, when the ball falls off the table, the only force acting on it is its weight *w* (neglecting, for now, the resistance of the air as it rushes past

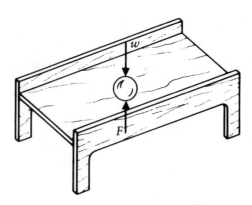

Figure 7.   The two forces acting on the rolling ball.

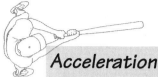

Acceleration

et us examine the method of calculating the acceleration experienced by the ball in Figure 5. The ball is initially moving at 80 mph to the east and later is moving at 80 mph to the southeast, that is, at an angle 45° below a horizontal line in the figure. We will assume that a period of time of 10 seconds has elapsed while the ball moved from position *A* to position *B*. The velocity at *B* is the vector **V**(B) and the velocity at *A* is the vector **V**(A). Since the average acceleration is the velocity difference divided by the time elapsed,

$$\mathbf{a} = \frac{\mathbf{V}(B) - \mathbf{V}(A)}{t(B) - t(A)}$$

The figure shows that the velocity difference is a vector with components $80(1 - \cos 45°) = 23.4$ mph to the left and $80 \sin 45° = 56.6$ mph downward. The magnitude of this acceleration can be determined by the use of the Pythagorean theorem, which states that the magnitude of the hypotenuse of a right triangle is equal to the square root of the sum of the squares of the sides, or $\sqrt{(23.4)^2 + (56.6)^2} = 61.2$ mph. The direction is specified by the angle θ shown. It is the angle whose tangent is 23.4/56.6 = 0.429, or θ = 23.2°. The acceleration is therefore 61.2 mph/s, or 6.12 mi/h/s in the direction 90° − 23.2°, or 66.8° south of due west.

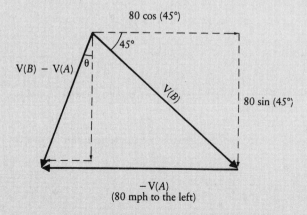

80 cos (45°)

45°

V(B) − V(A)

θ

V(B)

80 sin (45°)

−V(A)
(80 mph to the left)

Newton's Second Law

A mathematical description of Newton's second law of motion is

$$\mathbf{F} = m\mathbf{a}$$

The force **F** (a vector) is equal to the mass m of an object (a scalar) multiplied by its acceleration **a** (a vector). Since $m\mathbf{a}$ and **F** are both vectors and are equal, they must have both the same magnitude and the same direction. The force on the ball referred to in the box "Acceleration" must have been directed 66.8° south of due west. Its magnitude could be calculated simply by multiplying the mass of the ball by the magnitude of the acceleration.

the ball). In accordance with the second law, the downward speed of the ball begins to increase; that is, the ball experiences acceleration.

What is the amount of this acceleration? To find out, we turn to the sixteenth-century astronomer and experimental philosopher, Galileo Galilei. Galileo discovered, long before Newton's time, that the acceleration experienced by a body dropped from above the Earth is constant. (Actually, this value changes depending on the elevation of the body above the surface of the Earth, but for our purposes the variation is small.) This constant acceleration is given the special symbol g and is equal to 32 feet per second for each second the ball falls. For example, after 1 second the ball will be traveling at a downward speed of 32 feet per second. After another second, the ball's downward speed will have gained another 32 feet per second, so that its speed will be 64 feet per second. And so on.

Eventually, of course, the drag of the air rushing past the ball will produce a force equal and opposite to the weight of the ball, and, again in accordance with the first law, the ball will cease accelerating and fall at a constant velocity. This is, for example, the principle of a parachute. The greatly exposed area of the parachute's fabric serves to increase the drag force to a value high enough that a human can fall from great height and reach a "free fall" velocity small enough to prevent injury on landing. The free-fall velocity of a baseball in air is about 150 feet per second. As we shall see later, this aerodynamic drag force on a baseball is by no means unimportant, at least for long fly balls. For our present "thought experiments," however, let us agree to neglect it. We will add it to our analysis later, after our understanding of the basic laws of motion is more nearly complete.

Having analyzed the concepts of force (a push or a pull) and acceleration (the rate at which velocity increases), we can now set about understanding mass. As we have said, the downward force on a falling ball is its

weight w. Falling under the influence of gravity, the ball will accelerate at a rate given by g. Newton's second law tells us that the weight of the ball is equal to the mass of the ball multiplied by the acceleration due to gravity g:

$$w = mg$$

Far from the Earth or from any other body, the force of gravity becomes zero and an object becomes weightless. Since there are no forces on the object, its acceleration is also zero. Its mass, however, remains the same wherever the body is. If an object is placed on the surface of the Moon, it weighs about one sixth as much as it does on the surface of the Earth, because the acceleration of gravity on the Moon is one sixth that at the surface of the Earth. The mass is the same whether the object is on the Earth or the Moon or in outer space. It is equal to the weight of the object wherever it is, divided by the gravitational acceleration at that location:

$$m = \frac{w}{g}$$

According to Newton's second law of motion, an object under the influence of a constant force undergoes a constant acceleration that is directly proportional to the size of the force and inversely proportional to the mass of the object. Galileo guessed, and then verified experimentally, that the resulting motion of the object would be such that the distance traveled by the object would be proportional to the square of the time t. This result was later shown to be a direct result of Newton's laws of motion. The mathematical statement is

$$\text{Distance traveled} = \frac{1}{2} \, (\text{acceleration}) \, (\text{time}) \, (\text{time})$$

or

$$d = \frac{1}{2} \, at^2$$

Let us now return to our thought experiment about the ball rolling off the top of a table. Once the ball leaves the surface, it is influenced by a vertical force, the force of gravity or the weight of the ball. There is no horizontal force on the ball. There is a corollary to Newton's laws that tells us that as long as the horizontal and vertical forces on the ball are not dependent on one another, we can treat their effects separately. The total effect on the ball will be the combined effects of the vertical and horizontal forces. Suppose the ball rolls off the table at 50 feet per second without a horizontal force on the ball (remember: we are neglecting drag by the air). It will continue at a horizontal speed of 50 feet per second and, after 1 second, will be 50 feet from the edge of the surface. Its vertical (downward) displacement increases according to the preceding equation, where a is the acceleration due to gravity, 32 ft/s^2. After 1 second it will have fallen a distance $d = \frac{1}{2}(32)(1)^2 = 16$ feet. After 2 seconds, it will have traveled $2 \times 50 = 100$ feet horizontally and $\frac{1}{2}(32)(2)^2 = 64$ feet downward; after 3 seconds, $50 \times 3 = 150$ feet horizontally and $\frac{1}{2}(32)(3)^2 = 144$ feet downward, and so on.

Figure 8 illustrates what we have described. The first picture shows the horizontal motion alone, the middle picture shows the vertical motion alone, and the last picture shows the actual motion of the ball, a combination of the horizontal and vertical motions. Using vector arithmetic (see box, page 27), we find that the speed of the ball 1 second ($t = 1$) after falling from the table is $\sqrt{50^2 + 16^2}$, or about 52 feet per second in the direction shown.

We are now equipped to analyze the motion of a pitched or batted ball. Suppose that instead of rolling off a flat surface, the ball is thrown with a vertical velocity of 50 feet per second and also a horizontal velocity of 50 feet per second. If we were somehow able to turn off the force of gravity (or do our experiment on a space platform where there is no gravity), there would be no force on the ball, and it would travel along the straight dashed line shown in Figure 9. The speed of the ball would be $\sqrt{50^2 + 50^2}$, or about 71 feet per second, at an angle of 45 degrees from the horizontal. When gravity is present, the ball will fall below this line by the distances indicated in the figure and calculated from the equation $d = \frac{1}{2}at^2$ for $t = 1, 2, 3, 4$ and 5. After 1 second, it will be 16 feet below the dashed line. After 2 seconds, it will be 64 feet below the dashed line. Eventually, of course, the ball will hit the ground.

The curve described by the path of the ball shown in Figure 9 is called a *parabola* or *parabolic arc*. There is a simple mathematical expression for this arc. Let us agree to call the vertical distance from where the ball is thrown z, the horizontal distance x, the initial vertical velocity v_{z0}

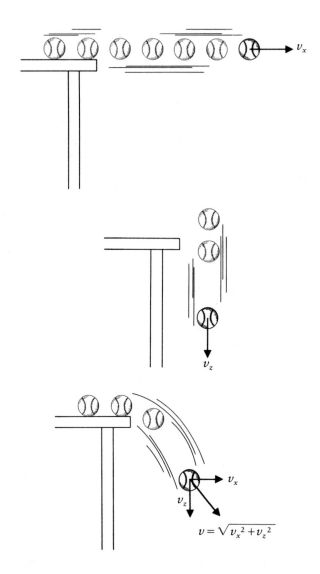

Figure 8. (Top) The motion of a moving ball if there were no gravitational force. (Middle) If the ball had no initial horizontal or vertical velocity, it would accelerate downward under the force of gravity (the weight of the ball) and in the last position shown have a velocity v_z. (Bottom) If the ball rolls off the table with some initial horizontal speed v_x the two effects described above act together.

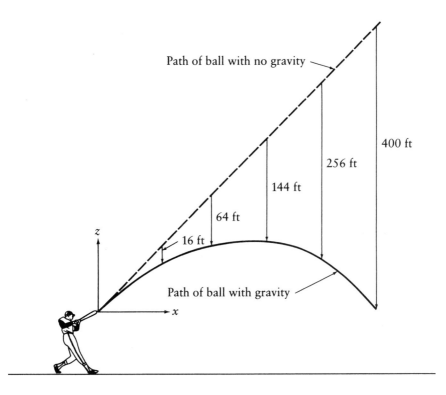

Path of ball with no gravity

400 ft

256 ft

144 ft

64 ft

16 ft

Path of ball with gravity

z

x

Figure 9. *A thrown or batted ball normally has both an initial vertical and an initial horizontal speed as it leaves the player's hand or bat. If there were no gravitational force and no air resistance, the ball would continue to move in its initial direction with the same speed (dashed lines). Under the force of gravity (but neglecting air resistance), the ball falls below this straight line at an accelerating rate (solid line). The actual path of the ball is a parabola lying in a vertical plane.*

and the initial horizontal velocity v_{x0}. The vertical distance, or height of the ball, at time t is just the difference between where it would be if gravity were absent, $z_1 = v_{z0}\,t$, and the distance through which it has fallen due to its weight when gravity is present, $z_2 = \frac{1}{2}\,gt^2$:

$$z = v_{z0}\,t - \frac{1}{2}\,gt^2$$

Equation for Range

hen the drag on a moving object due to the presence of air is negligibly small, its horizontal velocity remains constant. The equation $x = v_{x0}t$ tells us that the time required to move horizontally a distance x is the distance divided by the (constant) horizontal speed v_{x0}:

$$t = \frac{x}{v_{x0}}$$

This expression of time can be substituted for t in the equation $z = v_{z0}t - \frac{1}{2}gt^2$, with the result

$$z = v_{z0}\left(\frac{x}{v_{x0}}\right) - \frac{1}{2}g\left(\frac{x}{v_{x0}}\right)^2$$

Notice that when $x = 0$ in this equation, $z = 0$. This simply states that the ball was at the horizontal position $x = 0$ and the vertical position $z = 0$ at the same time. This was at time $t = 0$. At some time later $z = 0$ again. This occurs when the two terms on the right side of the equal sign in the above equation are equal to one another, or when

$$v_{z0}\left(\frac{x}{v_{x0}}\right) = \frac{1}{2}g\left(\frac{x}{y_{x0}}\right)^2$$

and this occurs when

$$x = \frac{2v_{x0}v_{z0}}{g}$$

This value of x is called the *range* (R) of the ball—the horizontal distance it will travel from the time it leaves the ground until it reaches the ground again.

The horizontal distance traveled by the ball at time t is

$$x = v_{x0}t$$

By solving the second equation above for t and substituting this into the first equation we can find the relation between z and x, from which we can derive an equation giving us the range R of the ball—that is, the horizontal distance the ball travels before returning to the height from which it was thrown. (See box, page 36.)

We now consider one last thought experiment with our fictitious ball. Until now, we have studied the motion of a ball that had an initial velocity and subsequently was subjected to a single force, the weight of the ball. The resulting motion of the ball could be visualized in a single plane, the x-z plane (see Figure 9). This motion is called *planar* or *two-dimensional motion*. We now imagine the motion of a ball when there is another force on the ball, in a direction mutually perpendicular to the x and z directions. We name this the y direction, and the motion is what we call *three-dimensional motion*. Imagine now a ball leaving the origin of the coordinates (the place where the three coordinate lines come together, that is, where x, y, and z are all 0). The ball has an initial velocity v_{z0} in the vertical (z) direction, an initial velocity v_{x0} in one horizontal (x) direction and no initial velocity in the other horizontal (y) direction. There is a vertical force, the weight of the ball, pulling downward in the z direction. As before, we can treat the three effects separately. The distances the ball travels in the z and x directions are unchanged from our previous thought experiment. However, the ball is now subjected to a force in the y direction, which we denote as F_y. Hence, according to Newton's second law, it will have an acceleration in the y direction of

$$a_y = \frac{F_y}{m}$$

and, since $m = w/g$, this acceleration can be expressed as

$$a_y = \frac{F_y}{w} g$$

According to the equation $d = \frac{1}{2} at^2$, it will move in the y direction as

$$y = \frac{1}{2}\left[\frac{F_y g}{w}\right] t^2$$

Equation $x = v_{x0} t$ can be solved for t and the equation above can be solved for t^2 and the results substituted into the equation so that $z = v_{z0} t - \frac{1}{2} gt^2$,

$$z = \left[\frac{v_{z0}}{v_{x0}}\right] x - \left[\frac{w}{F_y}\right] y$$

This is an important result. It is the equation of a plane. This plane is shown in Figure 10.

The Equation of a Plane and the Curve Ball's Path

What does the equation of a plane have to do with the existence of a curve ball? We first analyzed the trajectory of a ball that was subjected only to one force after it was thrown: a vertical (downward) force equal to the weight of the ball. The ball moved in a parabolic arc in a plane, the *x-z* plane. We have now shown that if the ball additionally experiences a horizontal force perpendicular to the *x-z* plane (that is, in the *y* direction in Figure 10), the ball still travels in a parabolic arc in a plane, but the plane, instead of being vertical, slopes in the *y* direction. *This can happen only if there is a horizontal force in the* y *direction.*

We have just described the path of a curve ball, of course. Figure 10 shows how a curve ball loops in the parabolic arc due to the force of gravity, but also moves in the y direction. Compare it with the pictures from *Life* magazine in Figure 1. The balls thrown by Carl Hubbell and shown in the figure are not moving in vertical planes, but in planes sloping to one side or the other. Clearly, there are horizontal forces on these pitches. These photographs were indeed the first concrete proof of the existence of the curve ball!

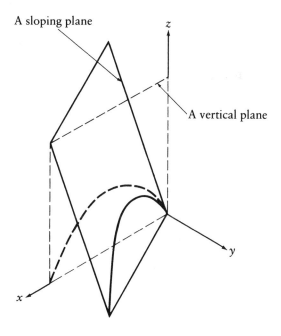

Figure 10. *Under the influence of an additional horizontal force perpendicular to the initial direction of travel, the ball will accelerate in this horizontal direction (the* y *direction in this figure). The ball moves in a parabolic arc but not in a vertical plane. Instead, it moves in a plane sloping downward and in the direction of the extra force. Comparison of this drawing with the* Life *magazine photograph (Figure 1) of Carl Hubbell's curve ball shows how that photograph proves Hubbell's pitch did indeed curve.*

The Laws of Motion and Pitched Baseballs

Let us now investigate some quantitative aspects of pitched baseballs and get accustomed to using the equations we have just derived. There are many interesting questions that can be answered before proceeding any further with theory.

Consider Dwight Gooden's fastball. Let us suppose it is thrown horizontally at 100 miles per hour and has no forces on it other than the downward pull of gravity. We will assume for simplicity that the ball travels 60 feet between the time it is thrown and the time it reaches home plate. (The distance to home plate from the pitcher's rubber is 60 feet 6 inches, but the ball only travels about 55 feet in the air because of the pitcher's arm and stride. We will use 60 feet throughout this book for consistency and simplicity.) We can compute how long the ball is in the air by using equation $x = v_{x0}t$. The ball travels a horizontal distance x at a horizontal speed v_{x0} in an amount of time

$$t = \frac{x}{v_{x0}}$$

We first express the speed in feet per second. Since there are 5,280 feet in a mile and 3,600 seconds in an hour, 100 miles per hour equals $(100)(5200)/(3600) = 147$ feet per second. The time required to reach the strike zone is therefore

$$t = \frac{60 \text{ ft}}{147 \text{ ft/s}} = 0.41 \text{ s}$$

The batter has only about four tenths of a second to decide whether the pitch will be in the strike zone and to swing his bat in such a way as to make an effective collision with the ball. Not many people can throw a baseball as fast as Dwight Gooden. An 85-mile-per-hour fastball has helped many a major league pitcher earn a good living. Such a pitch takes $(0.41)(100/85) = 0.48$ seconds to reach the strike zone. A 70 mile-an-hour fastball will be hammered out of the park by many high school players. Yet it reaches the strike zone in 0.58 seconds, only about one tenth of a second slower than a major league fastball. Ten to fifteen hundredths of a second is all that separates you and me from multimillion-dollar careers in major league baseball!

If a ball is thrown horizontally at 100 miles per hour (147 feet per second), how far will it fall vertically before crossing home plate? To do this computation, we can use equation $z = v_{z0}t - \frac{1}{2}gt^2$. Since the ball is thrown horizontally, $v_{z0} = 0$. The time required for the flight of the ball is 0.41 second, and the acceleration of gravity g is 32 feet per second per second. The vertical displacement is therefore

$$z = -\frac{1}{2}(32)(0.41)^2 = -2.69 \text{ ft}$$

or 2.69 feet downward. Even a ball thrown at the tremendous speed of a Dwight Gooden (or a Walter Johnson or a Bob Feller . . .) fastball drops by over 2.5 feet during its flight to home plate. In order to make a pitch rise, of course, some external upward force must operate on the ball. The magnitude of this force must be larger than the weight of the ball. Any upward force, however, would cause the ball to rise above the parabolic arc depicted in Figure 10. Since we are accustomed to seeing things move in these parabolic arcs, it seems entirely possible that departures from these arcs would be "seen" by a batter as a rise in the path of the ball. Or, maybe they do rise. Let us not decide this until we have had a chance to investigate the origins and magnitudes of forces that can be imposed on the ball by skillful pitchers.

For now, suppose a sideward force *can* somehow be put on the ball. How far will the ball curve on the way to home plate? First, to see what we mean by "how far," refer to Figure 11. This drawing is similar to that in Figure 10, except that a little more detail has been added. If we looked down from directly above the pitcher, and without depth perception, we would see the projection of the path of the ball in the x-y plane. By drawing vertical lines from the positions of the ball at several values of time, and marking where these lines cross the x-y plane, we can get an idea of the path we would see from our position above the pitcher. The ball is curving, or moving in the y direction, because there is a force in that direction. Its acceleration in the y direction is given by the last two equations on page 37. The distance the ball moves away from the x axis (the direction in which it was thrown) is given by the third equation. When we speak of how far a ball curves, it is the displacement in the y direction that we mean.

Given the horizontal speed of the pitch and the horizontal force on the ball, we can use the equation above to compute how far a ball will curve horizontally. A baseball weighs about 5.2 ounces. Suppose a horizontal force just half this large, 2.6 ounces, could be imposed on the ball. If the horizontal speed of the ball is 80 miles per hour (118 feet per second), the time to reach home plate is (from equation $x = v_{x0}t$) 0.512 second, and, by the equation above, the ball will curve horizontally by $y = \frac{1}{2}$ $\frac{1}{2}$ $(32)(0.512)^2 = 2.1$ feet. (The vertical distance that the pitch would drop due to its weight is $z = \frac{1}{2}$ $(32)(0.512)^2 = 4.19$ feet.) This would, of course, be quite a good pitch. In Chapter 3, we will see whether it is possible to impose a 2.6-ounce horizontal force on a baseball. Before going on to describe and discuss the forces that can cause a ball to curve,

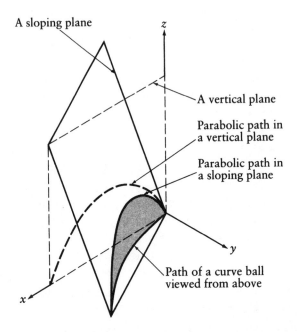

A sloping plane

z

A vertical plane

Parabolic path in a vertical plane

Parabolic path in a sloping plane

y

Path of a curve ball viewed from above

x

Figure 11. Most curve balls are thrown in such a way that the balls experience both a sideward force and a downward force (in addition to the force of gravity). A viewer directly above the pitcher and without depth perception would see the horizontal motion of the ball in the x-y plane as a parabolic arc in which the ball curves to the side.

it will be instructive to take one final look at the formulas we have been using.

The time it takes for a ball to go a horizontal distance x is given by the equation

$$t = \frac{x}{v_{x0}}$$

If we substitute this value of t into

$$y = \frac{1}{2}\left[\frac{F_y g}{w}\right] t^2$$

we find the following expression for the distance the ball curves

$$y = \frac{1}{2} \left[\frac{F_y}{w} \right] g \left[\frac{x}{v_{x0}} \right]^2$$

The form of this equation tells us much about the motion of a curving baseball. First of all, the distance y that the ball curves depends inversely on the mass of the ball: $m = w/g$. (As long as we confine our discussion to objects near the surface of the Earth, where g does not vary much, we can think of mass as reflecting the weight of the object.) A ball that is half as massive, that is, "heavy," as a baseball will curve twice as far; a ball that is twice as massive will curve only half as far. Ping-Pong balls, Whiffle balls, tennis balls, and the like can curve a lot even when subjected to small forces. This inverse relationship between the distance a ball curves and its mass is what prompted the author of the 1941 *Life* magazine article to write that "it is perfectly true that a person can put a curve on a Ping-Pong ball or a tennis ball . . . but a baseball is so heavy an object for its size in comparison to a tennis ball or Ping-Pong ball that the pitcher's spinning action appears to be insufficient to change its course."[8] We shall see.

Finally, the form of the equation just given can be used to shed light on the controversy about whether a curve ball breaks or curves smoothly. One popular explanation of the perceived break of a curve ball attributes it to an optical illusion. Imagine yourself standing by the side of a railroad track watching an oncoming train. To quote from an article in *Life* magazine (1953)[9]: "A railroad train seems to go faster and faster as it gets nearer a watcher until it appears to 'break' to the side with a great whoosh and whip by, causing the onlooker to duck—although the train's course and speed is constant. A batter is subjected to the same visual magnifications as he watches a ball speed towards him." But then why doesn't a straight ball appear to break? Clearly, this is not an adequate explanation. Let us look in a little more detail at the flight of a curve ball.

According to the above equation for the distance the ball curves, a ball subjected to a constant horizontal force curves in a smooth arc, but let us examine this arc more closely. Recall that we computed that an 80-mile-an-hour curve ball subjected to a horizontal force of 2.6 ounces would curve by 2.1 feet on its way to home plate. We could also obtain this directly from the equation, where we show y as a function of distance:

$$y(60) = \frac{1}{2} \left[\frac{2.6}{5.2} \right] (32) \left[\frac{60}{118} \right]^2 = 2.1 \text{ ft}$$

However, during the first half of its flight, it curves only

$$y(30) = \frac{1}{2}\left[\frac{2.6}{5.2}\right](32)\left[\frac{30}{118}\right]^2 = 0.52 \text{ ft}$$

This means it has to curve 1.58 feet during the last half of its flight, three times as far as it does during the first half. In fact, half of the total 2.1 feet that the ball moves in the y direction occurs during the last 17.6 feet of its flight toward home plate.

It is even more impressive to look at these figures in another way. During the first 10 feet the ball moves

$$y(10) = \frac{1}{2}\left[\frac{2.6}{5.2}\right](32)\left[\frac{10}{118}\right]^2 = 0.058 \text{ ft}$$

in the y direction (less than three quarters of an inch). During the first 50 feet, it moves

$$y(50) = \frac{1}{2}\left[\frac{2.6}{5.2}\right](32)\left[\frac{50}{118}\right]^2 = 1.45 \text{ ft}$$

in the y direction. Therefore, it curves by $2.1 - 1.45 = 0.65$ feet (nearly 8 inches) during the *last* 10 feet. The ball moves laterally more than 10 times further during the *last* 10 feet of the trip to the plate than it moves laterally during the *first* 10 feet. The ball is not simply drifting away from its initial direction along the x axis—it is *accelerating away* from the hitter. The parabolic arc* is smooth.

In this chapter we showed that if a horizontal force could be put on a baseball, the ball would follow a parabolic arc, that is, a curve. In the next chapter, we will explain where these forces come from. Then, in Chapter 4, we will describe several scientific experiments that were conducted to measure the horizontal forces on baseballs. The results of these experiments, together with the equations developed in this chapter, can then be used to calculate the deflections actually experienced by curve balls, knuckleballs and other pitches.

---

*Actually, the force is always perpendicular to the direction of motion of the ball, as we will see in Chapters 3 and 4. The motion of the ball under a force perpendicular to its motion is a circular arc.

Notes

1. W. F. Allman, "Pitching Rainbows," *Science,* October 1982, pp. 32–39.

2. Charles Einstein, ed., *The Fireside Book of Baseball,* Simon and Schuster, New York, 1987, p. 99.

3. R W. Madden, Letter to the editor, *The New Yorker,* May 24, 1941, pp. 53–54.

4. Noel Houston, Letter to the editor, *The New Yorker,* July 5, 1941, p. 44.

5. Lucy Stanton, Letter to the editor, *The New Yorker,* June 7, 1941, p. 70.

6. Robert Smith, *Baseball,* Simon & Schuster, New York, 1947, p. 73.

7. Ray D. Lillibridge, Letter to the editor, *The New Yorker,* June 7, 1941, p. 69.

8. "Baseball's Curve Balls: Are They Optical Illusions?" *Life,* September 15, 1941, pp. 83–89.

9. "Camera and Science Settle the Old Rhubarb about Baseball's Curveball," *Life,* July 27, 1953, pp. 104–107.

The Origin of Forces on a Baseball

The fact that tennis balls curve because of the spin imparted to them was noted as early as 1671 by Sir Isaac Newton.[1] In 1877 the English scientist Lord Rayleigh,[2] in a paper also describing the irregular flight of the tennis ball, credited the German engineer G. Magnus with the first explanation of the lateral deflection of a spinning ball. This phenomenon, in fact, derives its name from its explicator: the *Magnus effect*. Magnus's explanation, however, was published in 1853 (according to H. Barkla and L. Auchterlonie[3]), more than a century after a similar explanation had been given by Benjamin Robins[4] in *New Principles of Gunnery*. Both Magnus and Robins were interested in the trajectories of cannonballs and of musket balls. Robins experimented with the effects of spin on the curvature of the paths of musket balls by firing them from a gun with a slightly curved barrel. With the barrel bent to the left, the ball was forced into contact with the right side of the bore, thus imparting spin on the ball in the clockwise direction when viewed from above. The paths of the musket balls fired in this way were curved to the right.

47

A gun designed to throw curve balls was described in the June 1897 *Scientific American.*[5]

We present some engravings of the new gunpowder gun for pitching a baseball, tried at the Princeton ball field on June 8, 9 and 10. A charge of powder in a tube coiled about the barrel is ignited, the gases are delivered behind the ball and it is flung from the barrel. Two "fingers," thin plates of metal curved and covered with rubber, project over the thickness of the barrel, and impart a velocity of spin to the ball; this spin gives it a curved path.

The explanation given by Robins and Magnus went as follows: a spinning musket ball creates a whirlpool of air around it in addition to the

linear flow of the air past the ball due to its motion through space. This circulating air slows the flow of air past the ball on the left side and speeds it up on the right side. A theorem of Daniel Bernoulli, an eighteenth-century Swiss mathematician, states that when the speed of a fluid (like air) increases, its pressure decreases. (See box, pages 50–52) Thus, the left side of the ball, where the air speed is slower, experiences a higher pressure than the right side. The resulting pressure (and force) imbalance caused the ball to move laterally toward the right side.

What really happens is a bit more complicated than this. It involves a thin layer of fluid very close to the solid wall of the sphere called the *boundary layer*. Because every fluid possesses *viscosity*, there is a shearing force between adjacent regions of the fluid that move at different speeds. By rubbing your hands against each other or pushing a book across a tabletop, you can experience a shearing force. Let us see how shearing forces affect a fluid in motion.

Imagine a fluid flowing past a flat wall, and imagine that the fluid consists of many layers of molecules, one layer next to the other. Experiments have shown that the fluid layer immediately adjacent to the wall essentially "sticks" to the wall. Every other layer rubs against layers adjacent to it, resulting in shearing forces between the layers. Through these shearing forces, the faster-moving layers drag the slower layers along. The fluid far from the wall is practically unaffected by the presence of the wall. It moves at what we call the *free stream velocity*. The layer adjacent to the wall is totally affected by the presence of the wall in the sense that it "sticks" to the wall. There is a region between these extremes where the shearing force caused by the viscosity of the fluid is of great importance. This region, where the layers of fluid rub against one another to create a shearing force, is called the *boundary layer*, because it is present near solid boundaries. It is, of course, a group of layers in the sense that we have been using the term layer. Figure 12 is a drawing of motion in a boundary layer near a wall. Here the fluid is water. A pulse of electricity was sent through a tiny wire, the vertical line on the left side of the drawing, causing a chemical reaction that produced a colored line in the water. A photograph was taken a moment later. You can see that the colored water near the wall moved more slowly than that farther from the wall, indicating that the speed of the water decreased from the free stream value far from the wall to zero at the wall. The same phenomenon occurs when air flows past a baseball (or, equivalently, when a baseball moves through air). The

Bernoulli's Principle

ernoulli's principle states that regions of higher velocity in a fluid correspond to regions of lower pressure. This principle is actually a direct result of Newton's second law. To see this, consider what happens when a constant force F is applied to a mass m. The mass will accelerate and will attain a form of energy called *kinetic energy*. This energy results directly from work done on the mass by the force. The work done by a constant force is defined as the force multiplied by the distance through which it moves. According to Newton's second law, the work W is

$$W = F \cdot d = m \cdot a \cdot d$$

Now suppose the mass initially has zero speed. Then after a time t it will attain a certain speed v. The acceleration, a, is v/t and the distance the mass moves is the average speed multiplied by t, or $d = v_{ave}t = \frac{1}{2}vt$. The work can now be expressed as

$$W = m\left(\frac{v}{t}\right)\left(\frac{1}{2}vt\right) = \frac{1}{2}mv^2$$

The quantity $\frac{1}{2}mv^2$ is the kinetic energy.

Now consider the case of a fluid in motion. A line within the fluid to which the velocity vector is tangent is called a *streamline*. If the flow is steady, the fluid always moves along the streamlines. We then imagine a tube made up of streamlines. Since the flow is always tangent to the streamlines, the tube behaves as though it were a real tube in the sense that fluid flows through it without leaking from the sides. Such a stream tube is shown here.

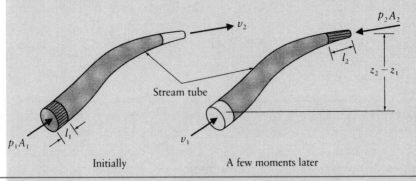

Stream tube

Initially · A few moments later

At a certain time, a certain identifiable portion of fluid occupies the shaded portion of the tube as shown in the left part of the figure. A moment later this fluid has moved to a new portion of the tube as shown on the right. The motion was caused by pressure forces. How much work has been done in moving the shaded portion of the fluid from the first position to the second position? The fluid within the large central portion of the streamline has undergone no change. It is as though the portion of fluid of volume $A_1 l_1$, located at the bottom left, the inlet of the stream tube, suddenly appeared at the upper right, the exit of the stream tube.

As long as no fluid is stored within the stream tube, the mass of fluid entering must be the same as that leaving. Using the symbol ρ to denote density (mass per volume), the mass entering or leaving is

$$\rho A_1 l_1 = \rho A_2 l_2$$

Work is done on the fluid by the pressure force $p_1 A_1$ in the amount $p_1 A_1 l_1$. Work is done at the exit of the stream tube in the amount $-p_2 A_2 l_2$. Note that this quantity is negative, because the directions of the force $p_2 A_2$ and the distance moved l_2 are opposite. Instead of work being done on the fluid, work has been done by the fluid. Work has also been done against the force of gravity in lifting the fluid. The gravitational force is the mass of the fluid times the gravitational acceleration, and so the work done by gravity is $-\rho A_1 l_1 g(z_2 - z_1)$. The kinetic energy increase caused by the work is the difference between $\frac{1}{2}mv^2$ of the portion leaving the tube and that entering the tube:

$$p_1 A_1 l_1 - p_2 A_2 l_2 - \rho A_1 l_1 g\, (z_2 - z_1) = \frac{1}{2}\, \rho A_1 l_1\, (v_2^2 - v_1^2)$$

If the density is constant, the volume flowing into the tube is the same as that flowing out, so that $A_1 l_1 = A_2 l_2$. In this case, the above equation can be divided by $-\rho A_1 l_1$, resulting in Bernoulli's equation

$$\frac{p_1}{\rho} + gz_1 + \frac{v_1^2}{2} = \frac{p_2}{\rho} + gz_2 + \frac{v_2^2}{2}$$

For the situations we will be studying (air at high speeds flowing past base-balls) gz_1, is very nearly equal to gz_2 and these two terms may be safely neglected. Furthermore, since positions 1 and 2 are arbitrary, $p/\rho + v^2/2$ must be the same everywhere along the streamline. Finally, we note that Bernoulli's equation is true when $v = 0$. Thus, one can theoretically always follow a streamline to a place where $v = 0$ and then switch to another streamline. We conclude that $p/\rho + v^2/2 + gz$ is constant throughout the fluid. Hence, if the speed, and therefore the kinetic energy, increases, the pressure must decrease so that the sum of the pressure and kinetic energy remains constant.

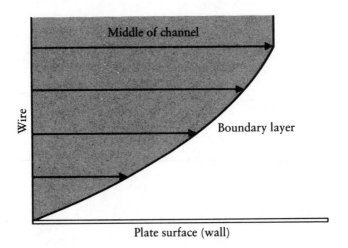

Figure 12.   *The behavior of a fluid flowing from left to right in a flat channel. At the bottom is the wall of a channel. The top of the drawing is the middle of the channel. The arrows pointing to the dark edge show how far the particles have gone since the dye was released. Particles far from the wall have moved with the free stream velocity, while particles closer to the wall (within the boundary layer) have moved more slowly.*

thickness of the boundary layer in that case is very small—only a millimeter or two. Yet, its presence is crucial to the ability of a pitcher to manipulate the paths of his pitches. Let us now see why.

If the viscosity of the air were zero, the air would flow around the ball as shown in Figure 13. Think of a small piece of fluid, called a *fluid particle*, within the fluid as a whole. As a fluid particle at point A in the figure approaches the ball and begins to move around it, from the front (at A) toward the equator at A'), it will accelerate. This means that its speed, and therefore its kinetic energy, or energy of motion, will increase. (See box, page 50.) In order for this to happen, the fluid particle must have had a force pushing on it. The force that causes the fluid particle to accelerate comes from a pressure difference between points A and A'. According to Bernoulli's principle, the pressure in the fluid at A is greater than the pressure at A'. Thus, in going from point A to point A', the kinetic energy of the fluid particle increases as a result of work done on the fluid particle by pressure force. Now imagine the same fluid particle

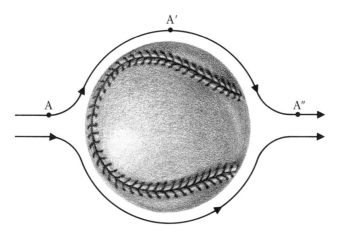

Figure 13.   *The flow of an ideal fluid having no viscosity. The speed of the fluid increases as it goes from point A to point A' and then decreases as it goes from point A' to A''. As the speed increases, the pressure of the fluid decreases because a pressure force is required to accelerate the fluid. The pressure decreases between points A' and A'' because a negative force is required to slow the fluid down. There is no net force on the sphere in this ideal situation because the pressure forces on the top and the bottom and on the front and back cancel each other.*

as it goes from point A' to point A'' near the rear of the ball. Again, because of Bernoulli's principle, exactly the reverse of what we have described above happens. The pressure at point A' is smaller than at A''; the fluid particle is slowed down, and the kinetic energy that was gained through work done by the pressure force between A and A' is given back to the rising pressure between A' and A''.

To understand the effect of shearing forces in the fluid flow past a ball, we first consider an easier example. Imagine yourself pushing a block across a frictionless table from A to A' as in Figure 14. As you push the block, its speed increases because of the work you do on the block. The pressure of your finger against the block is similar to the pressure of the fluid on the fluid particle in Figure 13. Now reverse the procedure as the block moves from A' to A''. As you push against the block in the opposite direction from its motion, the block slows again. (You have actually done negative work, because the block has done work on you!)

Now imagine yourself going through the same procedure with the block on a table that is not frictionless. Refer to Figure 15. This time, as you do work on the block to speed it up, you must work against a frictional force that wants to slow the block down. Still, you push hard enough to get the block up to a certain speed at A'. When you reverse the process after passing A', the frictional force is helping you to slow the block down. The block will, of course, stop before it comes to point A''. It will, in fact, reverse directions if you keep pushing.

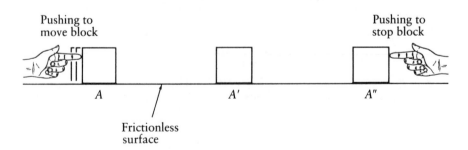

Figure 14. *The reversible nature of the force-acceleration interaction in the absence of friction. The block is initially stationary at point A. If we push the block along a frictionless table, it accelerates and reaches a certain speed at point A'. If we now reverse the force, the block decelerates until it comes to rest at point A".*

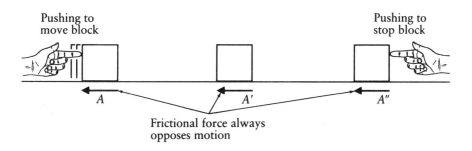

Figure 15. *If friction is present and the force profile of Figure 14 is used, the block will not reach point A″, because frictional forces always oppose motion. When a force toward the right is imposed at point A, the block accelerates towards point A′. Because of the frictional force, however, its speed will be lower than that of the block in Figure 14. When the negative, braking force is now imposed on the block, it is aided by the frictional force. The deceleration is faster, and the block comes to rest to the left of point A″. If the force continues to be imposed, the block actually reverses direction.*

Finally, imagine the (real) flow of air past a baseball with (viscous) frictional force included. Refer to Figure 16. The fluid particles in the boundary layer will behave similarly to the block on the frictional table in our thought experiment above. As a fluid particle in the boundary layer moves past point A' at the equator, it slows down. Of course, the free stream flow finally sweeps the fluid particle downstream. On the downstream side of the ball, however, the fluid particle stops before reaching the rearward pole. (See box, page 58.) A region called the *wake* region develops, where the air experiences a swirling, chaotic flow. You are probably familiar with wakes behind boats. You can also see this phenomenon on the downstream sides of bridge abutments. We call the point on the surface where the wake begins the *separation point*, and the phenomenon itself *boundary layer separation*.

Because of the chaotic flow, the pressure in the wake region is significantly lower than the pressure on the front of the ball. This is because the kinetic energy lost by the fluid particles in traveling from the equator to downstream is bound up in the turbulent eddies in the wake instead of being added back to the pressure. The existence of the wake creates a drag force. If the wake is symmetrical (which occurs when the ball is not spinning and when there are no irregularities on the ball's surface), the

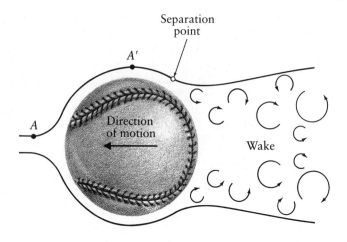

Separation
point

A'

A

Direction
of motion

Wake

Figure 16.   The flow of a real fluid (like air) around a bat. Since the fluid possesses viscosity, fluid particles in the boundary layer stop before going all the way around the ball. At some point beyond A' a region of backflow develops. The flow in the free stream ultimately pushes the fluid past the ball, but the development of the backflow near the ball produces a region of chaotic, swirling flow around the rear of the ball. This region is called the wake, and the point where it begins is called the separation point.

drag will simply slow the ball down. If, however, the wake is not symmetrical, it will have the additional effect of redirecting the flow coming off the rear of the ball.

Figure 17 shows a remarkable photograph of the flow around a spinning baseball taken by the late Professor F. N. M. Brown of Notre Dame University. The ball is spinning clockwise and the flow of the air is from left to right. Separation is delayed (occurs further back) on the top surface and occurs prematurely (further forward) on the bottom surface. Put rather simplistically, if the surface of the ball moves along with the flow, less energy is drained from the flow by the shearing forces of viscosity, and the separation is delayed—as happens with the top surface. If the surface of the ball moves opposite to the flow, the frictional effects are enhanced, and separation occurs earlier—as happens with the bottom surface. The shifted wake region has the effect of deflecting the flow of air downward as it passes the ball. This has the same effect on the ball

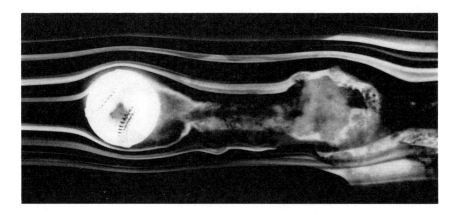

Figure 17.   A ball rotating in a wind tunnel. The ball is spinning clockwise and the air is flowing left to right. The paths of fluid particles of air are made visible by introducing streamers of smoke into the air. Since the upper surface of the ball moves in the same direction as the free stream flow, the air flows further around the ball before separation occurs. Conversely, the motion of the bottom surface of the ball is opposed to the flow, and separation occurs sooner. Because of the off-center wake, the flow of air past the ball is deflected downward in this case. The downward force required to deflect the wake must be balanced by an equal and opposite upward force on the ball by the air. (Photograph F. N. M. Brown, University of Notre Dame; courtesy, Thomas J. Mueller)

that a rudder has on a boat. If, as a boat moves through the water, the rudder is placed so that it deflects water toward the right, the rear of the boat is forced toward the left. The baseball in Figure 17 will be forced upward as the airflow is deflected downward. Clearly, by throwing a baseball with spin about one axis or another, a pitcher can cause the ball to curve to the right or to the left, or to rise above or fall below its normal parabolic arc.

If the ball has backspin, for example, it will rise above its normal parabolic arc. If the ball has topspin, it will drop. If it spins around a vertical axis like a top, it will move horizontally. Usually the axis of rotation is inclined. For example, a right-handed pitcher sees the curve ball leave his hand with its axis of rotation pointed to the upper left so that it has both topspin and a counterclockwise rotation (when viewed from above). This

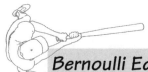

Bernoulli Equation with Loss

In a real fluid, another work term must be added to the analysis from which the Bernoulli equation was derived. This is the work done by a fluid particle in order to overcome viscous or frictional forces that are always present in any real fluid. Frictional forces do work to impede motion. Thus, the frictional term will serve to decrease the speed v_2 of the fluid at the end of the stream tube (as seen in the box on page 51) below the value predicted in that box. The Bernoulli equation with losses is

$$\frac{p_1}{\rho} + \frac{v_1^2}{2} + gz_1 = \frac{p_2}{\rho} + \frac{v_2^2}{2} + gz_2 + L$$

in which L is a positive quantity.

makes the ball drop (because of the topspin) and curve away from a right-handed batter (because of the counterclockwise rotation). A right-handed pitcher would see his screwball leave with its axis of rotation pointed to the upper right so that it has both topspin and a clockwise rotation. This ball drops and curves toward a right-handed batter (because of the clockwise rotation). The orientation of the axis of rotation depends on the pitcher's arm, wrist and finger actions during the delivery. For example, a fastball thrown with a straight overhand delivery will not curve to the side, whereas the same pitch thrown with a three-quarter overhand delivery will have some curve. Therefore, the direction of movement depends on the orientation of the spin axis. The amount of movement depends on the speed and the rotation rate of the pitch.

Table 1 shows typical values of these two parameters for a variety of pitches and pitchers. We do not use all the information in this table now, but we will repeatedly refer to it. These are "ballpark figures," not scientific data; they represent the melding of data from many sources. Of course, there are people whose abilities lie outside of these ranges, but they are atypical. The top five entries are for major league pitchers. Speed

Table 1 Typical Characteristics of Various Pitches

Type of Pitch	Speed (mph)	Typical Spin Rate (rpm)	Rotations En route to Plate
Fastball	85 to 95	1600	11
Slider	75 to 85	1700	13
Curve Ball	70 to 80	1900	16
Change-up	60 to 70	1500	14
Knuckleball	60 to 70	25 to 50	$\frac{1}{4}$ to $\frac{1}{2}$
Typical Fathers	40 to 50		
7- to 9-year-old boys	30 to 40		

refers to speed when the ball leaves the pitcher's hand, in miles per hour. The ball loses about 10 percent of its speed en route to the plate. Spin rate refers to the estimated spin rate in revolutions per minute; the range, for each of the top four lines, is about ± 300 rpm. Rotations refers to the number of times the ball turns around on its trip from the pitcher to the plate.

We have not discussed the pitcher's grip on the ball. The grip makes the difference between the fastball, the sinking fastball, the split-fingered fastball, the fork ball, the knuckleball, and the change-up, according to Jerry Kindall.[6] These pitches are all thrown with about the same arm motion, but the different grips produce different speeds, rotation rates, and axes of rotation. These three parameters in turn change the trajectories of the balls. In this book, we discuss these three parameters and not the grip on the ball.

Besides putting spin on the ball, there are other ways of manipulating the wake—some legal and some illegal. Pitches like knuckleballs, spitballs and scuff balls are other examples of pitches in which the pitcher manipulates the wake to produce a lateral force on the ball. Let us see how this manipulation is accomplished.

The German aerodynamicist C. Weisselsberger discovered early in this century that by placing a wire around the front part of a smooth sphere one could, under certain conditions, delay boundary layer separation and the formation of the wake (Figure 18). The wire apparently introduces turbulence into the boundary layer near the sphere, increasing its communication with the fluid further away from the surface, so that it can be swept further around the sphere. In other words, the shearing force between outer layers of the fluid is increased.

Because of the presence of its seams, a nonspinning baseball offers to an airstream a nonsymmetrical version of this picture. On one side of the baseball, the presence of stitches or seams may delay separation, while on the other side they may not. Figure 19 is a sketch showing what happens. With the seams in the position shown, the top part of the ball is smooth, but the seam region on the bottom causes the boundary layer to become

Figure 18.   The point where separation occurs and the wake forms can be manipulated by altering the surface of the sphere. (Permission of Office National d' Etudes et de Recherches Aerospatiales)

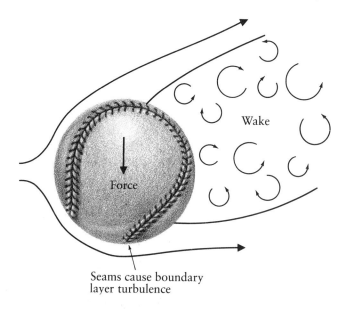

Wake

Force

Seams cause boundary
layer turbulence

Figure 19.   *The flow around a spinless ball. The ball is oriented in such a way that the top is smooth, while the stitches along the bottom surface act in much the same way as the wire in Figure 18. Separation is thus delayed on the lower surface, and the resulting upward deflection of the wake causes a downward force on the ball.*

turbulent, delaying separation. The wake shifts upward, redirecting airflow in that direction and forcing the ball downward. Note that again, as for the curve ball, it is the shifted wake that leads to the lateral force on the ball.

One reason the knuckleball is so notoriously difficult to hit is that there are an infinite number of ways to orient the seams. Each will cause a different force. No two pitches will behave in exactly the same way. Many people think that a knuckleball is a spinless pitch. Knuckleball pitchers know, however, that a good knuckleball is thrown so that it spins very slowly and makes perhaps one half a revolution on the way to home plate. As the locations of the seams relative to the airflow change, the position of the wake actually moves around, so the ball is forced first one way and then another. This explains the erratic paths of these pitches.

The knuckleball is normally thrown by holding the ball with the first knuckles of one or more fingers, rather than with the fingertips. As the ball is released, the fingers are extended, inhibiting the normal backspin of the ball.

There are a number of other ways to throw a baseball so that it has very little spin. One method is to make the fingers slippery by applying some foreign material like petroleum jelly or saliva to the fingertips, but not to the thumb. (Many of the old-timers chewed a substance called *slippery elm* for this purpose.) In releasing the ball, the pitcher squeezes it out between his fingers and thumb the way one would squeeze a watermelon seed, thus inhibiting backspin. As mentioned in Chapter 1, the knuckleball became common a few years after the spitball was outlawed. It was, in fact, first known as the dry spitter. It is interesting to note that the spitball was outlawed because it was said to be too difficult to control and therefore dangerous to hitters. According to Ray Miller, the spitball is, in fact, easier to control than is the knuckleball. There is other evidence that this is true. Many baseball insiders say that more major-league pitchers commonly throw spitballs than knuckleballs.

Another illegal pitch owing its behavior to shifting wake is the scuff ball or emery ball. The ball is prepared by scuffing it on one spot so that one side is much rougher than the other. The ball is then thrown so that the rough spot remains on one side. The rough spot delays boundary layer separation on that side, shifting the wake, and causing the ball to curve toward the roughened spot.

In this chapter, we have explained the lateral movement of a baseball in terms of altering the wake behind it. Using the analogy of a boat in water, we explained that pushing the tiller to the left moves the rudder to the right. This deflects the water (the wake) to the right, which pushes the rear of the boat to the left and turns the boat to the right. We showed two quite different mechanisms that could move the wake of a baseball. The first explains why a curve ball curves. The rotation of the curve ball delays the separation on one side and advances the separation point on the other, thus producing an off-center wake and, therefore, lateral motion of the ball. The second mechanism for moving the wake explained the movement of the knuckleball and scuff ball. A rough surface on one side of a nonspinning ball will delay the separation point and alter the wake, again producing a lateral force.

Now that we have seen how forces can be imposed on spinning and nonspinning balls, we turn to other questions. How large are these forces? How far can these forces make a ball curve? These are the topics of the next chapter.

Notes

1. Isaac Newton, "A Letter of Mr. Isaac Newton, of the University of Cambridge, Containing His New Theory About Light and Colors," *Philosophical Transactions of the Royal Society*, Vol. 7, 1671–1672, pp. 3075–3087.

2. Lord Rayleigh, "On the Irregular Flight of a Tennis-Ball," *Messenger of Math*, Vol. 7, 1877–1878, pp. 14–16.

3. H. Barkla and L. Auchterlonie, "The Magnus or Robins Effect on Rotating Spheres," *Journal of Fluid Mechanics*, Vol. 47, 1971, pp. 437–447.

4. Benjamin Robins, *New Principles of Gunnery*, Hutton, London, 1742.

5. "50, 100 and 150 Years Ago," *Scientific American*, June 1997, p. 12.

6. Jerry Kindall, *Sports Illustrated Baseball*, Harper & Row, New York, 1983.

The Flight of the Ball:

Forces on Curve Balls and Other Pitches

Since the photographic investigation published in 1941 by *Life* magazine, a number of articles have appeared in scientific journals and popular magazines seeking to answer definitively the question of whether a baseball curves—and whether it curves smoothly or "breaks." In April 1942, less than a year after the appearance of the *Life* magazine article, a note appeared in the *American Journal of Physics* describing an experimental study by F. Verwiebe[1] that was almost identical to the fictional one performed by Rick Rover. Five wooden frames with nets of fine cotton thread (rather than the wet tissue paper "used" in the Rover Boys experiment) were placed between the pitcher's mound and home plate. "Straight balls," "right-handed outdrops" and "left-handed outdrops" were thrown through the frames by pitchers on a college baseball team. The "outdrops" curved in the horizontal plane by 2.5 to 6.5 inches. The curve was reported to be smooth, with no sharp "break" observed. An interesting observation reported by Verwiebe was that the balls also

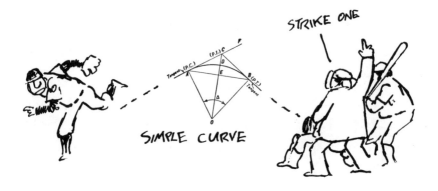

STRIKE ONE

SIMPLE CURVE

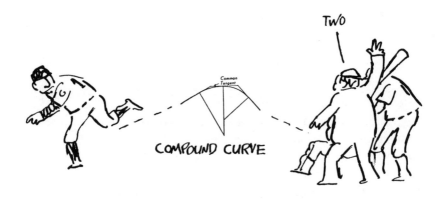

TWO

COMPOUND CURVE

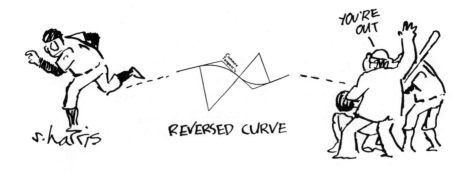

YOU'RE OUT

REVERSED CURVE

s.harris

"dropped more sharply than would be the case for free fall alone." We will have more to say about this feature of the curve ball later.

On June 19, 1949, *Look* magazine published "Visual Proof that a Baseball Curves," and on July 27, 1953, *Life* magazine, this time using Gjon Mili's strobe photography of pitches thrown by Ken Raffensberger of the Cincinnati Reds, reversed itself, claiming that the curve ball curves but does not "break." Some interesting new data came out of the photographic study. By painting one half of the baseball black, the investigators made it easy to see from the photographs how rapidly the baseball was rotating. The Raffensberger curve rotated at about 1400 revolutions per minute (rpm) and had a speed of 43 miles per hour (mph), which is very slow. (See Table 1.) Joseph Bicknell, a professor of aeronautical engineering at MIT, reportedly determined, using a wind tunnel test, that such a pitch should curve by approximately 6 inches.

In the summer of 1982, *Science 82* magazine, together with three scientists, Robert G. Watts of Tulane University, Jim Walton of General Motors, and Charlie Miller of MIT, set out to produce and study high-speed stroboscopic photographs of baseballs pitched by Scott McGregor, a pitcher for the Baltimore Orioles, and by Ray Miller, the Orioles's pitching coach. Watts and Walton[2] reported in an address to the American Association for the Advancement of Science that McGregor's curve balls did indeed curve, the total deflections being somewhat less than 1 foot.

These are not very impressive numbers. Any baseball player would reject them as being too small. Numbers that are considerably more impressive have, in fact, been reported. Watts and Walton reported that curve balls propelled by a pitching machine curved by about 2 feet. C. Selin[3] photographed and analyzed pitches thrown by college players in the Big Ten Conference and concluded that curve balls thrown by these pitchers suffered deflections of 0.4 to 2.2 feet. The speeds of these pitches were reported to range between 50 and 76 mph, while the balls rotated at 1150 to 2310 rpm.

Other wind tunnel tests (in addition to the one performed by Bicknell and referred to in the 1953 *Life* magazine article) have also been performed. L. J. Briggs[4] reported that an 80 mph curve ball spinning at 1800 rpm could curve as much as 17.5 inches, and Igor Sikorsky, a famous aerodynamicist, is reported by J. H. Drury[5] to have found that a baseball rotating at only 600 rpm and with a horizontal speed of 80 mph would curve by up to 19 inches.

In view of the differences in the deflections of curve balls reported by different writers, it might prove both interesting and useful to examine the original sources a bit more closely in order to see whether the different results are really at odds with one another, or whether there is a way to reconcile them.

We begin by pointing out that many people think of a curve ball curving only horizontally, and we have perhaps served to perpetuate that myth through our example involving a horizontal force at the end of Chapter 2. What Verwiebe measured, for example, were the horizontal deflections of what he referred to as "right-handed and left-handed outdrops." The fact that the vertical deviation of a curve ball from its otherwise parabolic trajectory is often larger than its horizontal deflection is well-known to those who play baseball professionally, and even to many serious amateurs. In fact, one customarily sees and hears a word like "outdrop" when pitchers describe their pitches. We might recall that the 1941 *Life* magazine article referred to Cy Blanton's curve ball as being "a straight ball with a decided drop," and that Verwiebe observed in his 1942 experiment that the right-handed and left-handed outdrops "dropped more sharply than would be the case for free fall alone." Ray Miller pointed out that this is by design, not by accident. In his 1981 article, Bill Allman[6] reported:

"The Ball has to break down," says Ray Miller, pitching coach for the Baltimore Orioles, "or it's not an effective pitch. The hitting area of the bat is three inches wide and eight inches long. A ball that breaks horizontally is just going to hit another part of the bat."

A baseball bat is essentially a cylindrical piece of wood that is used to strike the pitched ball. It is generally swung in such a way that the axis of the cylinder is approximately parallel to the ground (that is, approximately horizontal). An error in the horizontal position of the bat is therefore nowhere near as serious as an error in its vertical position. Missing by an inch or two horizontally is often not terribly serious, but an error of a fraction of an inch in the bat's vertical position can mean the difference between a line drive and a pop-up. As P. Kirkpatrick[7] rather humorously pointed out, "The 1962 world championship was finally determined by an otherwise perfect swing of a bat which came to the collision 1 mm too high to effect the transfer of title." We will suggest in the last section of Chapter 5 that the error was probably larger than this. However, the argument that the margin for error is greater horizontally than vertically remains true.

The forces that give rise to the "outdrop" have the same origin as the forces on the curve ball that we described in Chapter 3. They are caused by the spin of the ball. The direction of the force on a moving, spinning ball is determined by the relative directions of the ball speed (or the opposite of the air flowing past the ball) and the axis of rotation of the ball. In fact, the lateral (curving) force is in a direction mutually perpendicular to the direction in which the ball is traveling and to the direction of the axis of rotation. If the pitcher throws a sidearm curve, he can probably make the axis of rotation of the ball nearly vertical, so that the lateral forces will be horizontal. Generally, pitchers throw with a motion somewhere between directly overarm and sidearm. The result is that the axis of rotation is not vertical. Figure 20 shows the result. The lateral

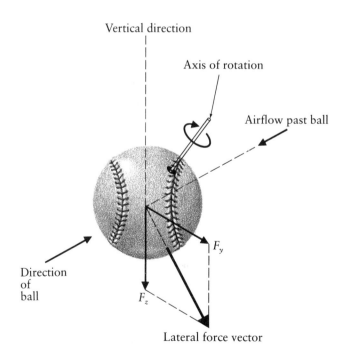

Vertical direction

Axis of rotation

Airflow past ball

F_y

Direction of ball

F_z

Lateral force vector

Figure 20   *The force on rotating ball is mutually perpendicular to the axis of rotation and the direction of air flow past the ball.*

force is not horizontal, and can be represented (remember the discussion of Figure 2) as two forces F_y (horizontal) and F_z (vertical).

Since Verwiebe did not give quantitative information about the "drop" part of his "outdrops," it is impossible to tell just how much the curve balls of his experiments were deflected from the parabolic path caused by gravity alone acting on the balls. Suppose, however, that the pitcher is able to cause a force F_y to act horizontally on the ball and, in addition, to cause a force of F_z to act downward on the ball. In a sense, the ball would then have an apparent weight of $w + F_z$. The weight of the ball causes it to fall through a distance $z_2 = \frac{1}{2} gt^2$. The extra downward force causes the ball to drop by an additional distance

$$z_3 = \frac{F_z g t^2}{2w}$$

This derivation is similar to that of the equation

$$y = \frac{1}{2} \left[\frac{F_y g}{w} \right] t^2$$

which is the expression of horizontal displacement due to a force F_y. The two displacement values caused only by the forces F_y and F_z are perpendicular to each other. To find either the total force or the total displacement, we use the same method that we used to calculate the speed v in Figure 7 from the individual components v_y and v_z. Thus, the nongravitational force acting on the ball is $F = \sqrt{F_y^2 + F_z^2}$, and the deflection from the parabolic path of the equation

$$z = v_{z0}t - \frac{1}{2} gt^2$$

is

$$d = \sqrt{z_3^2 + y^2} = \frac{F g t^2}{2w}$$

If, as Ray Miller's remarks suggest, the balls photographed by Verwiebe curved downward more than they curved horizontally, the z_3 could be a good deal larger than y. In any case, the values he measured and reported were certainly smaller than the real deflection of the curve.

What about Scott McGregor's curve ball? The deflection of McGregor's curve ball of somewhat less than 1 foot as reported by Watts and Walton was the total deflection; that is, both horizontal and vertical departures from the parabolic gravitational trajectory were considered. However, this experiment was flawed because the night before the photography session, McGregor had been called upon to pitch for the Orioles. Therefore, he was quite understandably throwing with somewhat less than his maximum effort in the photographic session. The measured deflection cannot be expected to reflect those of a major league curve ball.

The 1953 *Life* magazine article reported that the "widest variance" of the Raffensberger curve ball "is in mid-flight . . . where it measured 6 inches." Elsewhere in the article, it is reported that "recreating as nearly as possible Raffensberger's pitch in a wind tunnel with an ingenious gadget, Bicknell demonstrated that the ball was thrust sideways approximately six inches." The implication is that Bicknell measured the force on a baseball spinning at 1400 rpm in a wind tunnel with air speed set at 43 mph (the measured rotation and speed rate of Raffensberger's curve ball, respectively). He must then have computed the deflection of the ball using an equation like this

$$y = \frac{1}{2}\left[\frac{F_y g}{w}\right] t^2$$

Interestingly enough, his calculation seems to agree almost exactly with the "variance" of 6 inches actually measured. However, his 6-inch deflection and that reported by *Life* probably refer to different things. The one calculated by Bicknell refers, in all probability, to the displacement of the immediately preceding equation. The 6-inch "variance" of the Raffensberger curve ball most definitely does not, as we will now show.

The curved line in Figure 21 depicts the path of a curve ball as viewed directly from above, that is, in the x-y plane. (You may want to look back at the discussion of Figure 11 in Chapter 2.) The authors of the *Life* article joined "the beginning and the end of the curve's arc" with a straight line and measured the distance between the curve's arc and a straight line at the midpoint. Some geometric distances are shown in the figure. The letter d is the distance the pitch travels in the x direction, y is

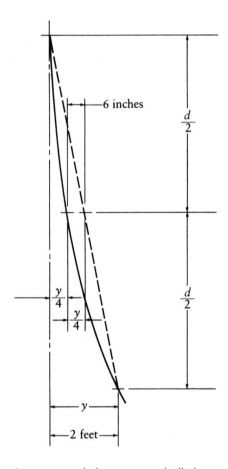

Figure 21 *When Life magazine reported that a curve ball thrown by Ken Raffensberger curved by 6 inches, they were referring to widest variance between the path of the ball in midflight and a straight line connecting the initial and final position of the baseball. In the text, we show that this is actually only one fourth of the actual deflection of the ball. The actual horizontal deflection of the ball, measured from the line it would have taken without the force caused by its spin, was an impressive 2 feet.*

the total deflection. Halfway along its path (at $d/2$) the straight line path would have moved over by $y/2$, but the parabolic curve would have moved only $y/(2)^2 = y/4$. The difference, $y/4$, represents the 6-inch measurement reported. Evidently, the horizontal deflection of the Raffensberger curve ball was an impressive $y = 4 \times 6$ inches $= 24$ inches! It is quite likely that Raffensberger threw a sidearm curve, a pitch that exaggerates the horizontal motion at the expense of the vertical motion, because he knew this was what the *Life* magazine people were going to measure. It is difficult to be sure why Bicknell reported a deflection of only 6 inches, because his measurement method was not reported in sufficient detail.

Does the Amount of Curve Depend on Velocity or Velocity Squared?

The article by Drury states that Sikorsky found that the deflection y of a spinning baseball on its path toward home plate is directly proportional to the rotation rate ω, the square of the velocity v of the ball, and the square of the time t required to reach home plate, and inversely proportional to the mass m of the ball:

$$ y = \frac{K\,\omega v^2 t^2}{m} $$

Until recently, the most systematic experiment on the forces of spinning baseballs that is reported in useful detail in the scientific literature is one conducted by L. J. Briggs of the National Bureau of Standards. The results of Briggs's tests indicate, as did the Sikorsky results, that the lateral force on the baseball is proportional to the rotation rate multiplied by the square of the velocity. However, these results conflict with a theorem in fluid mechanics called the Kutta-Zhukovskii theorem, which states that whenever a spinning two-dimensional object is moving through a fluid, and there is a net circulation of the fluid about the object, there results a force mutually perpendicular to the direction of the motion of the object and the direction of the axis of the spin. This force has a magnitude that is proportional to the product of the velocity and the rotation rate ωv. Experimental measurements of lift on rotating cylinders appear to bear this out. It seems reasonable that the lift forces on a rotating sphere would also depend on ωv rather than ωv^2.

One of us (Watts) and R. Ferrer have conducted a careful series of experiments designed to determine which of these is true.[8] The experimental arrangement that we used consisted of a subsonic wind tunnel, a device for measuring lift on a spinning ball, and devices for measuring the

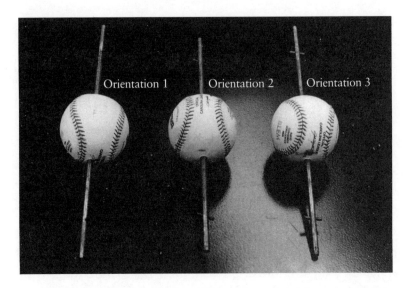

Figure 22   *In order to measure the forces on spinning baseballs, and to determine the effect of the orientation of the seams relative to the spin axis, we mounted three baseballs on thin shafts with three different seam orientations as shown in this figure, and studied them in a wind tunnel.*

rotational and free stream velocities. Three baseballs were impaled on 6.3-mm diameter shafts with the seams in the positions shown in Figure 22. The shaft of a particular ball was then mounted in a frame as shown in Figure 23. An impeller mounted on the end of the shaft was used for creating rotation by using a high-speed air jet issuing from a nozzle. The entire device was mounted in the test section of a subsonic wind tunnel. Air in the tunnel flows from right to left in Figure 23.

Our measurements showed that the force on the baseball is proportional to ωv, not to ωv^2. The equation for the force on a spinning baseball that we obtained from our data is

$$F_L = 6.4 \times 10^{-7}\, \omega v$$

where F_L is in pounds, v is in feet per second, and ω is in revolutions per minute. Using this equation, we find that an 80-mph curve ball rotating at 2000 rpm would experience a lateral force of 0.15 lb, or 2.4 oz. which

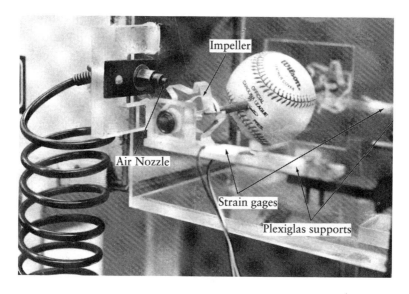

Figure 23   *The baseballs in our experiment were mounted in the student laboratory wind tunnel at Tulane University as shown in this figure. Air from a nozzle was used to induce rotation and the wind tunnel forced air past the ball. The rotation rate was measured with a strobotachometer. Forces on the ball induced forces in Plexiglas supports. Strain gauges in the supports were used to infer the forces on the spinning ball.*

is just about what we said we needed back in Chapter 2. Figure 24 illustrates this equation graphically.

The reason we mounted the balls with three different seam orientations was to test whether this is an important factor in how far a curve ball curves. Many baseball players believe that it is an important factor, and that orientation 1 in Figure 22 would result in a better curve than orientation 2 because orientation 1 puts "more strings into the wind." We found the orientation of the seams had virtually no effect on the force. Certain grips may, of course, allow the pitcher to get a higher rotation rate. As far as the force by the air on a rotating pitch is concerned, however, the positions of the seams are apparently of little significance. What is very important is the orientation of the axis of rotation of the ball. To maximize the curve, the axis of rotation should be precisely perpendicular to the direction of motion of the ball. Otherwise, the lateral force

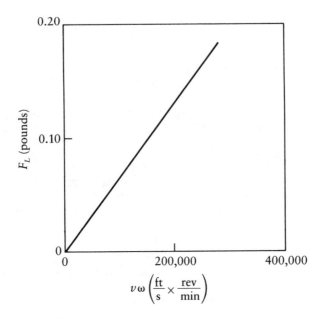

Figure 24   Our measurements show that the lift forces on a spinning baseball are almost linearly dependent on the product of the velocity and the rotation rate. While our measurements were limited to air speed values less than 40 miles per hour, because of limitations of the wind tunnel, the use of dimensionless groups, as explained in the text, allows us to use data by other authors to infer that this is also true at higher air speeds.

caused by the rotation of the ball will decrease approximately as the sine of the angle between the axis of rotation and the direction of motion.

Both the time-honored Kutta-Zhukovskii theorem and the results of our experiments cast doubt on the velocity-squared results of Briggs and Sikorsky. Our measurements were performed in a wind tunnel in which the maximum speed of the air was about 40 miles per hour. Pitched baseballs travel as fast as 100 miles per hour. There comes the natural question, therefore, of whether our results are valid at speeds more than twice as large as those used in our experiments. Measurements of the forces on spinning golf balls by P. W. Bearman and J. K. Harvey[9] and by William Gobush (private communication) at high velocities also indicate that F_L depends on ωv, not ωv^2. In fact, if the difference in surface areas of the two types of balls is taken into account, the results of Bearman and

Harvey and those of Gobush are very nearly identical to our results (see the immediately preceding equation).

This raises an important point. How does the lift force depend on the surface area of the ball? It makes sense also that the lift force should depend on other things, like the density and the viscosity of the air flowing past the ball.

Dimensionless Groups and the Lift Coefficient

From a scientific point of view, the immediately preceding equation is sloppy. The dimensions of quantities on either side of the equal sign in an equation must match. Since the force is in pounds, v is in feet per second, and ω in revolutions per minute, the constant 6.4×10^{-7} of the equation must have units of

$$\frac{\text{pounds} \times \text{seconds} \times \text{minutes}}{\text{feet} \times \text{revolution}}$$

Note that this constant does not have units related to properties of the ball or air density—properties, one would expect, affect the lift force. In general, this indicates that something is missing from the formulation. For example, as we indicated above, if we performed our experiments with a different-sized ball or in a different fluid, the "constant" should have a different value. We can get around this problem by the appropriate use of dimensionless groups. Let us begin our illustration of this idea with a somewhat far-fetched example of a familiar idea.

Suppose Galileo had performed the experiment in which he verified that the distance an object falls near the surface of the Earth is proportional to the square of the time elapsed. He would have measured

$$d = -16t^2$$

Had he the opportunity to perform his experiment on the Moon, he would have found that (approximately)

$$d = -2.6t^2$$

This is, of course, because the multiplier in front of t^2 is not really a constant. It has dimensions of length divided by time squared. Its value differs because not only the gravitational constant is different for the Earth and the Moon but, perhaps more important, its value is different when different units of length and time are used.

Notice that we can regroup the equation $d = \frac{1}{2}at^2$ (with $a = -g$) as

$$\frac{d}{gt^2} = -\frac{1}{2}$$

The value of the constant on the right-hand side is $-\frac{1}{2}$ and it is truly constant, independent of whether the experiment is performed on Earth or on the Moon and independent of the units used. It is dimensionless, as is the *dimensionless group* d/gt^2. Scientists like to cluster their variables in dimensionless groups, especially when they are trying to make sense of experimental data. The constants that emerge are then true constants. Obviously, the variables that form the groups must have consistent units so that the groups themselves are dimensionless. For example, if d is in feet and g is in feet per second per second, then t must be measured in seconds.

It turns out that to fully describe the results of the lift force experiments three dimensionless groups are required. They are conventionally written as

$$Group\ 1\ =\ C_L\ =\ \frac{2F_L}{\rho A v^2}$$

$$Group\ 2\ =\ SP\ =\ \frac{R\omega}{v}$$

$$Group\ 3\ =\ Re\ =\ \frac{2vR}{\nu}$$

The quantity A in Group 1 is the cross-sectional area of the ball, R in Group 2 is the radius of the ball, ρ in Group 1 is the mass density of the air, and ν in Group 3 is kinematic viscosity of the fluid. The quantities A and R are obviously properties of the ball. Some clarifying comments are in order concerning ρ, the air density. It is common in many places to give the density of a gas or liquid in pounds per cubic foot. This is weight density. To get the mass density ρ, this must be divided by g. This is because mass is weight divided by g. A second comment is that $R\omega/v$ must be dimensionless. Since R has dimensions of length and v length per time, ω must have units of inverse time $(1/\text{time})$. We have previously used revolutions per minute for this quantity. To change this to $1/s$, we note that one revolution is 2π radians (a radian is a dimensionless angular measure) and 1 minute is 60 seconds. Hence,

$$1\,\frac{\text{rev}}{\text{min}} \times \frac{2\pi\,\text{rad/rev}}{60\,\text{s/min}} = \frac{\pi\,\text{rad}}{30\,\text{s}}$$

To get radians per second from revolutions per minute, multiply by π and divide by 30. For example,

$$2000\,\frac{\text{rev}}{\text{min}} = (2000) \times \frac{\pi}{30} = 209\,\text{rad/s}$$

The three dimensionless groups are related to one another in such a way that Group 1, which is called the *lift coefficient* and is denoted C_L, is a function of Group 2, which is called the *spin parameter* and is denoted *SP*. It has been shown experimentally that unless the ball is very smooth (much smoother than a baseball) C_L is not affected by Group 3. For higher speeds or for very smooth balls, like billiard balls, C_L would also be a function of Group 3, which is called the *Reynolds number*, after the British scientist Osborn Reynolds. It is written in the abbreviated form *Re*.

To show how well this works, we have plotted data points from several experiments performed by different people in different laboratories with different types of balls (Figure 25). Watts and Ferrer used real baseballs, Gobush used real golf balls, and Bearman and Harvey used a dimpled ball somewhat like a golf ball but with a diameter 2.5-times-larger.

For a typical curve ball, ω is at most 2000 rev/min, or 209 rad/ sec. Using a velocity of 85 mi/h (125 ft/s) and the radius of a baseball ($R = 0.119$ ft), we see that *SP* has a typical value of $(0.119)\,(209)/\,125 = 0.2$. We expect that it would seldom be greater. When *SP* is less than about

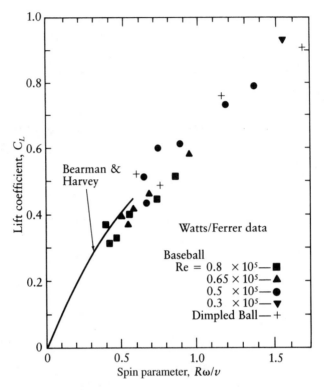

Figure 25 *This figure shows lift data plotted in the way a scientist would plot it. The coordinates are the lift coefficient, C_L, and the spin parameter, SP, as explained in the text. Watts and Ferrer's data are plotted along with that of Gobush and that of Bearman and Harvey. Typical values of SP for baseballs range between 0.1 and 0.2. In this range, C_L varies almost linearly with SP.*

0.4, the relation between C_L and SP is almost a straight line of slope unity. Thus a good approximation is $C_L \approx SP$, or

$$\frac{2F_L}{\rho A v^2} = \frac{R\omega}{v}$$

Replacing the cross-sectional area A with πR^2 and rearranging yields

$$F_L = \frac{1}{2}\rho\pi R^3 \omega v$$

We can now see how the "constant" in the equation $F_L = 6.4 \times 10^{-7} \omega v$ is related to the size of the ball and the density of the air. We can also see that the lift force varies directly with the product ωv.

As a final observation about dimensionless groups, we note how the data of Bearman and Harvey and the data of Gobush imply that we can use our results with confidence at larger ball speeds than we used in our experiments. Gobush tested balls at values of Re up to 2×10^5, and Bearman and Harvey as high as 2.38×10^5. Comparison of these values implies that our data are valid to speeds at least up to $(2.38/0.8) = 2.98$ times our maximum wind tunnel speeds, or up to 119 miles per hour.*

Now getting back to baseball, if we assume an 80-mph curve ball rotates at 209 rad/s, then using the equation $F_L = \frac{1}{2} \rho \pi R^3 \omega v$ we can calculate

$$F_L = \frac{1}{2}(0.0023)\,\pi\,(0.0017)\,(209)\,(117) = 0.15\ \text{lbs} = 2.4\ \text{ounces}$$

which is in agreement with the equation derived using our original measurements.

Nonspinning or Slowly Spinning Baseballs: The Knuckle Ball

Systematic measurements of the forces on nonspinning baseballs have also been made. In 1968, Richard Gonzalez, a mechanical engineering student at Tulane University, measured the forces on a nonspinning baseball in a wind tunnel and reported the results in his bachelor's thesis. Two years later the measurements were carefully repeated and extended. The results were published in the *American Journal of Physics*

---

*Those readers who are intimately familiar with the Magnus effect—the phenomenon leading to the lift force on a spinning sphere—will realize that the physics can be more complicated than our explanation might lead one to believe. Under certain conditions the lift coefficient can depend strongly on the dimensionless group Re, which is called the *Reynolds number*. The lift can, under certain conditions, actually reverse in direction, for example. It turns out, however, that such strange phenomena, which are probably related to transition to turbulence of the boundary layer flow, only occur on smooth spheres. The seams on a baseball and the dimples on a golf ball make the balls much too rough for Magnus reversal to occur. While this is a fascinating topic deserving of scientific scrutiny, it is not relevant to our considerations here.

by Robert G. Watts and E. Sawyer.[10] Watts and Sawyer mounted a baseball on a device that could be used to measure the forces on the ball. The ball was then placed in a wind tunnel, initially in the position shown in Figure 26. With the ball oriented in this way, a drag force could be measured (in the direction of the wind velocity), but there was no force in the direction perpendicular to the wind direction (F_L in Figure 26). The ball was then placed in different orientations by rotating it about the vertical axis (varying the angle ϑ in the figure). As expected, the nonsymmetrical locations of the seams caused a variation in the position of the wake, and the result was a sideward force F_L. Figure 27 shows how the magnitude of this force varied with the seams in different positions for a wind velocity of 68 feet per second (46 miles per hour). The force changes cyclically as the orientation is changed. A ball thrown with an orientation 30 degrees from the position shown in Figure 26 would be pushed to the left (from the pitcher's perspective) with a force of about 0.1 pound. A similar pitch with an orientation 60 degrees from that shown in Figure 26 would be pushed to the right with a force of about −0.1 pound. Thus, a pitch

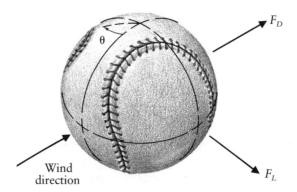

Figure 26   To measure the forces on a spinless baseball, Watts and Sawyer mounted a ball in a wind tunnel in the position shown. The ball is shown oriented in a symmetric position relative to the position of the seams. The angle θ is equal to zero. The ball was impaled on a thin vertical rod through its center. By turning it about the vertical axis, the seams could be oriented in various nonsymmetrical positions, and the forces measured by strain gauges in the supports in much the same way that the spinning ball forces were measured.

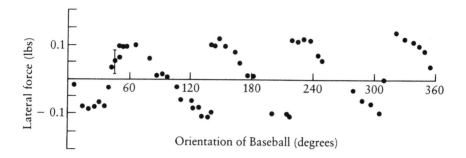

Figure 27 *The lateral (or sideward) force varies strongly with the orientation of the seams. A nonsymmetric location of the seams affects the flow in the boundary layer near the ball differently on different sides of the ball. This causes the wake to shift on one direction or another. When the wake shifts in one direction, a force is imparted on the ball. The data shown here were made at a wind speed of 46 miles per hour.*

thrown without spin could curve to the right or to the left. In fact, if a pitcher was unlucky enough to throw a spinless pitch with the ball oriented perfectly symmetrically, it would not curve at all!

Obviously, it is best if the pitch can be thrown with just a little rotation, because the direction of the force will change in midflight. This is why the knuckleball's path is erratic. Figure 28 illustrates two cases, a knuckleball thrown so that it rotates a quarter turn on the way to home plate, and one that rotates a half turn. The first begins curving toward the left and reverses itself. The second reverses itself three times. The more slowly rotating pitch has a maximum deflection that is much larger than that of the more rapidly rotating pitch. Remember that when a force is applied to a mass, it takes a little time to get the mass moving because it has to accelerate. The velocity does not change immediately. If the force changes direction too rapidly, the mass simply never gets anywhere. It turns out that the maximum displacement of a slowly rotating ball is

$$d_{max} = \frac{g \times F_{L\,max}}{w\omega^2}$$

where $F_{L\,max}$ is the maximum force in pounds, w is the weight of the

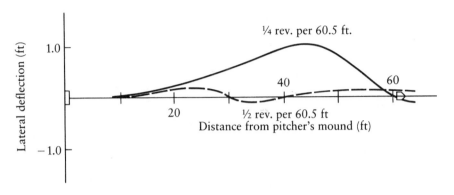

Figure 28   *If a pitch is thrown so that it spins very slowly, the lateral force on the ball will change in magnitude, and perhaps even in direction. The ball whose path is shown with a solid line was moving at 46 miles per hour and it rotated 90 degrees on the way to home plate. The lateral force, initially zero, increases to a magnitude of almost 0.1 pound. It then decreases in magnitude, reverses direction and, reaches about 0.1 pound in the other direction. If the ball rotates 180 degrees as shown with the dashed line, it reverses direction three times. Remember, though, that a force produces only an acceleration—a gradual change in velocity rather than an immediate change. If the force reverses too often, the ball does not have time to move very far. A good knuckleball must rotate a little, but not too much.*

ball in pounds, and ω is the rotation rate in radians per second. This equation works only when the ball rotates at least one quarter revolution on its path to home plate. At rotation rates higher than about 2 or 3 revolutions on the way to home plate, a different phenomenon occurs, namely, that already described in our discussion of curve balls. When the ball rotates less than one quarter revolution, the force goes through less than one cycle (see Figure 27), and the above equation again cannot be used.

The lateral force shown in Figure 27 was obtained with a wind speed of 68 feet per second. However, Figure 29 gives an indication of how the maximum magnitude of the force, F_{Lmax} changes with the speed of the pitch (or the wind speed). This figure shows that the mag-

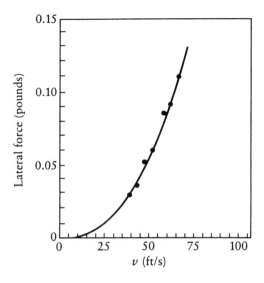

Figure 29   *The maximum lateral force on the spinless ball varies approximately with the square of the speed. If the maximum force on the ball as presented in Figure 27 at 46 miles per hour is 0.1 pound, the force on a typical 70 miles per hour knuckleball is $(\frac{70}{46})^2$ (0.1) = 0.23 pound. Since it is moving faster, however, it has less time to move, and the distance it moves laterally will be the same as that shown in Figure 28. This is explained in the text.*

nitude of the lateral force on the knuckleball varies as the square of the speed. Its maximum absolute value is about 0.1 pound (i.e., varying from about 0.1 pound to the left to 0.1 pound to the right) when v is 68 feet per second. A major league pitcher might be expected to throw a knuckleball a little faster, approximately 100 to 110 feet per second (70 to 75 miles per hour). At these speeds, the magnitude of the maximum lateral force is 0.23 to 0.27 pound. An equation for the magnitude of the maximum lateral force is

$$F_{L\max} = 2.16 \times 10^{-5}\, v^2$$

where $F_{L\max}$ is in pounds and v is in feet per second. In dimensionless terms, the lift coefficient C_L is constant with a value of 0.42. Hence, according to our dimensionless group arguments, since $C_L = 0.42$ the above equation can be written in the form

$$F_{L\text{max}} = 0.21 \, \rho \, Av^2$$

Scuff Balls

One of the authors (Watts) performed a set of experiments to measure the force on a "scuff ball." The cover of the ball was roughened by scratching it with a bottle top in one spot. This ball is shown in Figure 30. It was mounted in the wind tunnel so that the scuffed spot was on one side of the ball. It was not rotated. The lateral force on the ball in the direction of the scuffed spot is shown in Figure 31. Its value varied with the square of the wind speed as follows:

$$F_L = 1 \times 10^{-5} \, v^2$$

Figure 30   *Scuff balls are prepared by scratching one part of the ball with a bottle cap, belt buckle or whatever the pitcher can sneak to the mound. The ball is then thrown in such a way that the scuffed spot is at the axis of rotation of the ball. As the air flows around the ball the scuffed spot causes flow separation to be delayed, shifting the wake toward the side opposite the scuffed spot and inducing a force in the direction of the scuffed spot.*

where F_L is in pounds and v is in feet per second. The lift coefficient is again constant, this time with a value of 0.194. Hence, for this particular scuff ball

$$F_{Lmax} = 0.097 \, \rho \, Av^2$$

How Far Do They Curve?

We have now discovered that the lateral forces on rapidly spinning and slowly spinning balls are similar in magnitude to those we used in some simple trajectory calculations in Chapter 2. Let us now explore some consequences of the way these forces vary with the speed of the pitch and the rotation rate of the ball. First, consider the curve ball.

The data (that of Watts and Ferrer and that of Bearman and Harvey) indicated that the lateral force on a curve ball is proportional to the speed of the pitch times the rotation rate. Thus, we can say that

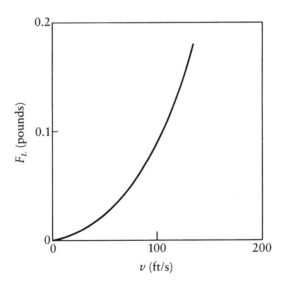

Figure 31   The force on a scuff ball may be larger the more extreme the scuffing. Data for one particular ball are shown here. The force increases as the square of the velocity, reaching a value of about 0.17 pound at 90 miles per hour.

$$F_L = C\,\omega v$$

where the constant C, according to the equation

$$\frac{2F_L}{\rho A v^2} = \frac{R\omega}{v}$$

is $\rho\pi R^3/2$. On the other hand, the equation

$$y = \frac{1}{2}\left[\frac{F_y\,g}{w}\right]t^2$$

tells us that such a force will move the ball in the lateral direction by a distance

$$d = \frac{F_L\,gt^2}{2w} = \frac{C\,\omega vgt^2}{2w}$$

Now, t is the time required for the ball to reach home plate, that is, the distance D to home plate divided by the speed v of the pitch. Thus,

$$t^2 = \frac{D^2}{V^2}$$

so that, from the equation

$$d = \frac{F_L\,gt^2}{2w} = \frac{C\,\omega vgt^2}{2w}$$

the total deflection of the curve ball is

$$d = \frac{\rho\pi R^3 \omega gD^2}{4wv}$$

Consistent units must be used in this equation of course; i.e., d in ft. ρ in lb-s/ft^4, R^3 in ft^3, ω in rad/s, g in ft/s^2, D^2 in ft^2, w in lb, and v in ft/s. This equation is illustrated in Figure 32. This equation shows that

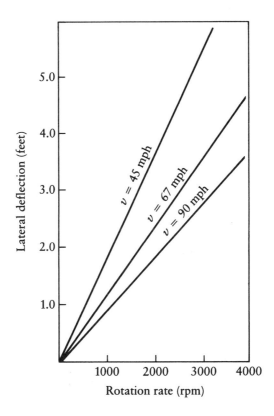

Figure 32   An illustration of the equation $d = (\rho\pi R^3 \omega g D^2)/(4wv)$.
The lateral force on a curve ball is directly proportional to the speed of
the pitch and the square of the time to reach home plate. Since the time
required to reach home plate is inversely proportional to the velocity, a
slower pitch (at a given spin rate) curves further than a faster pitch. Of
course, the batter has more time to look at the ball when the pitch is
slower.

the faster the ball spins, that is, the higher ω, the further it will curve.
Also, the slower the pitch, the further it curves. However, the slower
pitch with more curve may be easier to hit than a fast pitch with less
curve, because the batter has more time to track it and predict its trajectory.

Next, consider the knuckleball. Let us suppose the ball is thrown so
that it rotates N times on the way to home plate. The rotation rate is N

revolutions per 60 feet multiplied by v feet per second multiplied by 60 seconds per minute, or

$$\omega\, \frac{\text{rad}}{\text{s}} = \frac{N\,\text{rev}}{60\,\text{feet}} \times \frac{v\,\text{feet}}{\text{s}} \times \frac{2\pi\,\text{rad}}{\text{rev}} = \frac{Nv\,\pi}{30}\, \frac{\text{rad}}{\text{s}}$$

Thus, from the equation for maximum displacement,

$$d_{\text{max}} = 2920\, \frac{F_{L\text{mass}}}{w(Nv)^2}$$

The value of $F_{L\text{max}}$ was given earlier. Combining that formula with the immediately preceding equation, we can see that

$$d_{\text{max}} = \frac{6.3 \times 10^{-2}}{wN^2}$$

If $N = 1/2$ and $w = 0.32$ pounds, then d_{max}, $= 0.8$ feet. This equation is illustrated in Figure 33. The immediately preceding equation is not valid for $N < 0.25$. However, J. W. Hollenberg[11] has run computer simulations of knuckleballs with very low spin rates, for example, 3 rpm, and has found typical lateral displacements of 2 feet.

The distance the knuckleball moves is independent of velocity, a remarkable result. It shows that a fast knuckleball will be deflected by the same distance as a slow knuckleball. The deflection of a nearly spinless pitch is independent of the speed of the pitch. This may imply that a fast knuckleball is better than a slow knuckleball, because it will be deflected by the same distance but in a shorter period of time.

The same is true of the scuff ball. According to the equation

$$y = \frac{1}{2} \left[\frac{F_y\, g}{w} \right] t^2$$

$$d = \frac{F_L\, g t^2}{2w}$$

If we substitute F_L from the equation $F_L = 10^{-5}\, v^2$ and the time t to reach

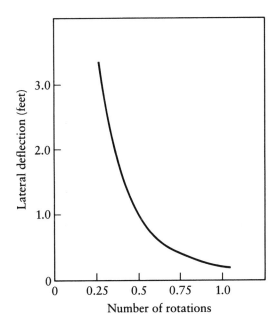

Figure 33 An illustration of the equation $d_{max} = (6.3 \times 10^{-2}) / (wN^2)$. Since the force on a knuckleball depends on the velocity squared, the lateral deflection is independent of the speed of the pitch. It depends only on the rotation rate. Balls rotating by less than about 45 degrees, one eighth of a revolution, on the way to home plate will generally move in only one direction (that is, there will be no reversals). The larger the rotation rate of the ball, the more direction reversals will occur, but the smaller the maximum lateral distance the ball will move.

home plate from equation $t^2 = D^2/v^2$, we find that for the particular scuffed ball used in our experiments,

$$d = 1.73 \text{ feet}$$

The distance the ball is deflected probably depends upon the size and roughness of the scuffed patch. However, again this distance does not depend on the speed of the pitch.

We can summarize what we have learned as a set of graphs. Figure 32 is a graphical picture of the equation concerning the deflection of a curve ball:

$$d = \frac{\rho \pi R^3 \, \omega g \, D^2}{4wv}$$

Read in the rotation rate of the ball along the bottom, or horizontal, axis. Trace along a vertical line until you reach the straight line giving the speed of the pitch. Now go horizontally straight to the vertical axis on the left and read the deflection of the pitch.

Figure 33 is the same sort of thing for a knuckleball, except this time the maximum deflection does not depend on the speed of the pitch. Read in the number of rotations on the way to the plate along the horizontal axis at the bottom. Trace a line up to the curve and then across to the left-hand vertical axis for the maximum deflection.

The Rising Fastball

In Chapter 3 we posed the question "Can a fastball rise?," as Walter Johnson's was reported to have done. Let us see what our measurements tell us. In order to rise, the ball must have enough backspin to induce a lift force larger than the weight of the ball, 5.125 ounces, or 0.32 pound. If we suppose that the ball is thrown at 147 feet per second (100 miles per hour), we can use the equation $F_L = 6.4 \times 10^{-7} \, \omega v$ to compute the required backspin. The result is

$$\omega = \frac{0.32 \text{ pound}}{(6.4 \times 10^{-7})(147 \text{ feet per second})} = 3400 \text{ rpm}$$

The maximum rotation rate of balls thrown by college baseball players reported by Selin was 2310 revolutions per minute. Can Nolan Ryan or Dwight Gooden make the ball spin 1100 revolutions per minute faster? Perhaps someday, someone will have the opportunity to photograph a Ryan or Gooden fastball and count the revolutions. We think it is very unlikely.

We have spent the last three chapters discussing the pitcher's art and science. Yet, pitchers are not the only players who manipulate the spin of the baseball to their advantage. According to Bill Shirley of the *Los Angeles Times*, Steve Garvey says the ball carries better if you undercut it a little. Clearly, what Garvey is proposing to do is put backspin on the ball to give it some lift. It is obvious that if the ball is undercut too much it will be popped up or fouled back. To understand how much is too much, and many other interesting things about the hitter's art and science, we need to turn our attention from the mound to the batter's box, as we do in the next chapter.

Notes

1. F. Verwiebe, "Does a Baseball Curve?," *American Journal of Physics*, vol. 10, April 1942, pp. 119–120. A scientist's view.

2. Robert G. Watts and Jim Walton, "Gravity's Rainbows: The Kinematics of Baseballs and Bodies," Annual Meeting of the American Association for the Advancement of Science, Detroit, May 1982. A speech on baseball, gymnastics and diving.

3. C. Selin, "An Analysis of the Aerodynamics of Pitched Baseballs," *The Research Quarterly*, vol. 30, no. 2, 1959, pp. 232–240. High-speed pictures capture baseball spin.

4. L. J. Briggs, "Effect of Spin and Speed on the Lateral Deflection (Curve) of the Baseball; and the Magnus Effect for Smooth Spheres," *American Journal of Physics*, vol. 27, 1959, pp. 589–596. Measurements of the forces on spinning balls.

5. J. H. Drury, "The Hell It Don't Curve," in *The Fireside Book of Baseball*, Charles E. Einstein, ed., Simon and Schuster, New York, 1956. Baseball stories.

6. William F. Allman, "Dance of the Curve," *Science 81*, vol. 2, October 1981, pp. 90–91. Comments by scientists and players about curve balls.

7. P. Kirkpatrick, "Batting the Ball." *American Journal of Physics*, vol. 31, 1963, pp. 606–613. A physicist looks at batting.

8. Robert G. Watts and R. Ferrer, "The Lateral Force on a Spinning Sphere," *American Journal of Physics*, 1987. Careful measurements of the forces on spinning balls.

9. P. W. Bearman and J. K. Harvey, "Golf Ball Aerodynamics," *Aeronauts Quarterly*, vol. 27, 1976, p. 112. Careful scientific measurements of the forces on spinning golf balls.

10. Robert G. Watts and E. Sawyer, "Aerodynamics of a Knuckleball," *American Journal of Physics*, vol. 43, 1975, pp. 960–963. A scientific account about why knuckleballs flutter and jump erratically.

11. J. W. Hollenberg, "Secrets of the Knuckleball," *The Bent of Tau Beta Pi*, vol. 77, no. 4, 1986, pp. 26–30. Measurements and computer calculations of knuckleball motion.

Bat Meets Ball:

Collisions

It has been said many times that batting a baseball is the single most difficult act in all of sports.[1] At any given moment only a half-dozen players in the combined major leagues are accomplishing the task of getting a base hit as often as once in every three attempts. Paul Kirkpatrick,[2] a physicist at Stanford University, pointed out that this is at least partly attributable to the fact that in the attempt to hit a baseball solidly with a bat there are so many things that can go wrong. Let us suppose that for a given pitch there is an ideal way for the bat to meet the ball in a collision. This means that there is an ideal location of the bat at the instant that the collision occurs. The location of the bat is determined by three variables: its height above the ground, its position along a line between home plate and the pitcher's mound and its position in the direction across home plate. There is also an ideal angular position of the bat (three more variables), an ideal speed (three more, one component in each direction), and an ideal angular speed (still three more). Obviously, if the batter adjusts all these things perfectly, but swings too early or too late, it would do his cause no good. Thus, he can also fail in his task through the mismanagement of time. Taken together with the fact that any of the above

95

13 variables can be missed on either the high side or the low side, this means that there are 26 ways to fail.

Not all the variables are equally important, of course. A baseball bat is a cylindrical object, and its correct position when it strikes a baseball is approximately horizontal and perpendicular to the path of the approaching ball. A small error in the rotational position of the bat about a vertical axis may cause an otherwise perfectly hit home run to go into the left or right field bleachers instead of the center field bleachers, while a larger error would transform a home run into a long but harmless foul ball. On the other hand, an error in the rotational position about the axis (length) of the bat is of no importance at all. The most critical variable is the vertical position of the bat. Conventional wisdom is that a very tiny error here (a few millimeters) can spell the difference between success and failure. However, we will scrutinize this few millimeters assumption in the last section of this chapter and again in Chapter 6.

An interesting aspect of the game of baseball is the fact that the batter, the offensive player, is not essentially in control of the ball. This is a situation quite different from that in any other sport played with a ball. In baseball, the defensive team controls the ball perhaps 95 percent of the time. The hitter's job is to hit the ball in such a way that the defensive team loses control long enough for him or one of his teammates to circle the bases. A team scores not so much by controlling the ball as by forcing his opponents to lose control. The ultimate method of doing this is to hit the ball in such a way that it goes out of the playing field (from fair territory) as a home run. In a sense, the batter has momentarily gained full control of the ball and the defensive team has completely lost control. Even as the ball is thrown toward the batter's box, the pitcher has a certain amount of control over the path of the ball. The batter's task of causing the defensive team to lose control is made more difficult because the pitcher can manipulate the path of the ball in a way presumably known only to himself and the catcher. In addition to imposing a variety of forces on the ball by both legal and illegal means, most pitchers expose batters to some added psychological stress by occasionally reminding them how much a baseball thrown at 85 to 95 miles per hour hurts when it hits them. Most batters already know this, of course, so the reminder can be indirect. Ryne Duren, a flame-throwing relief pitcher for the New York Yankees during the 1960s, sometimes warmed up by throwing several pitches over the catcher's head. He would then remove his thick glasses and, while wiping them with his handkerchief, squint at the batter taking his place in the batter's box. Don Drysdale, the former hard-throwing right-hander for the Los Angeles Dodgers, is said to have once quipped, "The pitcher has to find out if the hitter is timid. And if the hitter is timid, the pitcher has to remind the hitter that he's timid." Thus, another variable, fear (ballplayers use the term *respect*), enters the batter's formula for failure.

Hitting a baseball can be thought of as consisting of five interrelated parts: (1) seeing the ball as it approaches the plate, (2) deciding when and where to swing the bat, (3) actually swinging the bat, (4) the collision between the ball and the bat and (5) the path of the ball after it leaves the bat. In this chapter, we will discuss only the fourth of these, the collision between the bat (a modified right circular cylinder) and the ball (a sphere). In Chapter 6 we will discuss the path or trajectory of the ball after it leaves the bat. In Chapter 7, we will study the head and eye movements of a batter trying to follow a pitch.

Momentum, Impulse, and Collisions

In order to explain the collision process, we must introduce another idea from physics: the concept of momentum. The momentum principle stems directly from Newton's second law:

$$F = ma$$

Suppose a constant force F is applied to an object of mass m so that its speed increases from v_0 to v over a time interval t. Since the average acceleration is the change in speed divided by the change in time,

$$F = m\frac{v - v_0}{t}$$

or, after rearranging the terms,

$$Ft = mv - mv_0$$

The quantity mv is the *momentum* of the object; mv_0 is its initial momentum, and mv is its momentum after a force F has been applied to it for a time t. The quantity Ft is called the *impulse*. The impulse on an object during a collision is equal to the change of momentum resulting from the collision. If there is no force, the momentum remains unchanged.

Consider now what happens when two objects (a bat and a ball) collide head on. To simplify matters, we will assume the ball and bat velocity vectors are colinear, as shown in Figure 34. Assume the ball has mass m_1 and speed v_{1b}, and the bat has mass m_2 and speed v_{2b} before the collision. Since the collision is head on, v_{1b} and v_{2b} are in opposite directions. This is indicated in the figure by placing a minus sign in front of the ball's velocity, v_{1b}. In a typical collision of bat and ball, the force is very large and the time period over which the forces act is very small. These forces greatly deform one or both objects. However, the deformations cannot be seen by the naked eye because they happen so fast. Figure 35 is a picture of a baseball-bat collision taken with a high-speed camera. The baseball, being considerably softer than the bat, is deformed to a surprising degree during the collision. It is obviously subjected to a very large force. The force can be as large as 1500 pounds for a period as short as 0.002 second. After this time interval, the force between the ball and bat

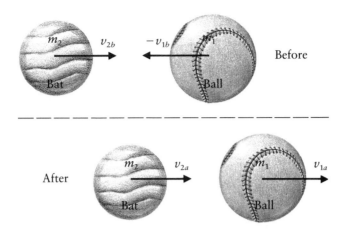

Before

After

Figure 34   A collision between a ball (on the right and moving toward the left in the top diagram) and a bat (on the left and moving toward the right). The subscript 1 is for the ball, 2 is for the bat. The subscript b is for before the bat-ball collision (the top diagram), and a is for after the collision (the bottom diagram). (From A. T. Bahill and W. J. Karnavas, Biological Cybernetics, 62: 89–97, 1989.)

returns to 0 and the ball springs back to its nearly spherical shape. According to another law of motion, Newton's third law, if the force imposed by the mass m_2 on the mass m_1 during the collision is F_1, then the force imposed by m_1 on m_2 is $-F_1$. Using F_1 in the equation for the principle of momentum, the impulse equation gives

$$F_1 t = m_1 v_{1a} - m_1 v_{1b}$$

Since the ball is moving to the left, v_{1b} is a negative number. Similarly, applying the momentum principle to the mass m_2 gives

$$-F_1 t = m_2 v_{2a} - m_2 v_{2b}$$

Adding the right- and left-hand sides of these equations and rearranging produces the *conservation of momentum equation*:

$$m_1 v_{1b} + m_2 v_{2b} = m_1 v_{1a} + m_2 v_{2a}$$

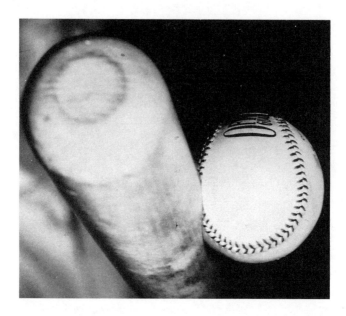

Figure 35   A high-speed flash photograph of a ball striking a baseball. Notice the deformation of the ball, indicating the enormous magnitude of the impulsive force at this instant. (Courtesy Harold E. Edgerton, Massachusetts Institute of Technology, Cambridge, Mass.)

The *total* momentum of the two objects taken together is the same before and after the collision. When you think about it, this is not surprising. Each object exerts a force on the other, but these forces cancel, and there is no net force on the two objects when taken together. The equation represents the law of the conservation of momentum: the total momentum must remain unchanged, when there is no externally applied force.

In the collision between a baseball and a bat, there is another force, the force that the batter's hands exert on the bat. If, however, it were anywhere near as large as the force acting between the bat and the ball, it would sting the batter's hands badly. This might happen, for example, when the ball hits the bat close to the batter's fists. Usually (if the ball is hit properly) the effect of the force from the batter's hands on the collision is small, and the immediately preceding equation is accurate. There is another problem, however. If the collision between the ball and the bat does not occur at the bat's center of mass, the collision will induce rotational motion in the bat. We will consider such collisions in the last section of this chapter. For now, we will consider only the velocities of the centers of mass of the bat and ball.

The conservation of momentum equation above tells us a lot about head-on collisions, but it cannot tell us all we want to know. Usually in collision problems we know the masses m_1 and m_2 and the speeds v_{1b} and v_{2b} of the two objects before the collision. We wish to determine the speeds v_{1a} and v_{2a} of the objects after the collision. There are two unknowns. We cannot find them both from only one equation. To determine either of the speeds after the collision we need another equation.

We need an equation that describes the physical characteristics of the two bodies that are colliding. Clearly, it is going to make a great deal of difference whether object m_1 is a baseball or a lump of clay, and whether m_2 is a wooden bat or a foam-rubber pillow. In other words, the "bounciness" of both objects is involved. A baseball will bounce higher when dropped on a concrete floor than will a softball or a lump of clay. None of these will bounce high when dropped on a thickly carpeted floor.

The property that scientists use to describe the "bounciness" associated with the collision between two objects is the coefficient of restitution e. Its value is

$$e = \frac{\text{relative speed after collision}}{\text{relative speed before collision}}$$

In our case,

$$e = -\frac{v_{1a} - v_{2a}}{v_{1b} - v_{2b}}$$

To get some feeling for what this equation tells us, let us construct a simple example. Suppose we drop an object (of mass m_1) on a concrete floor (of mass m_2). In this special case both v_{2b} and v_{2a} are obviously zero (the concrete floor does not move either before or after the collision) and

$$e = -\frac{v_{1a}}{v_{1b}}$$

Suppose the object strikes the floor at -10 feet per second (negative because it is going down). If the ball-floor combination were perfectly "bouncy," $e = 1$, the floor would just reverse the ball's direction, and v_{1a} would be $+10$ feet per second. Conversely, a ball of clay would not bounce at all. In that case, $e = 0$. The baseball-bat collision lies somewhere between these two extremes. The manufacture of baseballs is carefully controlled to ensure that the coefficient of restitution is 0.55. From the definition of the coefficient of restitution and from the equations of Chapter 2, it can be seen that if you drop a professional league baseball onto a concrete floor, it will rebound e^2 of the original distance; for example, if you drop it from 6 feet, it will rebound about 22 inches. The coefficient of restitution depends on the velocities of the objects before the collision,[3] but these effects are small so we will use a constant value of e throughout most of this book.* Assuming e to be constant also makes the mathematics much simpler. The coefficient of restitution for the collision of two objects depends on properties of both objects. However, for a bat-ball collision most of the potential variability of the coefficient of restitution is due to the ball. So people often ignore the bat and speak of the coefficient of restitution of the ball.

The Standard Major League Baseball

A baseball is currently required to weigh between 5 and 5.25 ounces and to have a diameter between 2.86 and 2.94 inches. Before 1973, major league baseballs had been for many years assembled in the Spalding factory in Chicopee, Massachusetts. Then the work was done for a few years in Haiti, and since 1976 in Taiwan. The balls are constructed in layers. At the center is a composition cork ball encased in two thin layers of rubber, one black and one red. On this is tightly wound 121 yards of blue-gray wool yarn, 45 yards of white wool yarn, another 53 yards of blue-gray wool yarn, 150 yards of fine cotton yarn, and a coat of rubber cement. It is enclosed in a cowhide cover hand-stitched with 216 red cotton stitches.

The weight and diameter of a baseball have been standardized since 1872, when machines for the manufacture of balls came into use. However, this did not solve the problem of standardizing the coefficient of

*Our computer programs use the following equations for the coefficient of restitution (Coe): for a CU31 aluminum bat and a softball Coe = 1.17 $(0.56 - 0.01(v_{1b} - v_{2b}))$, and for a wooden bat and a hardball we use Coe = 1.17 $(0.61 - 0.001(v_{1b} - v_{2b}))$ where the collision speed is in mph.

restitution of the balls. A debate over this elusive physical property has been going on at least since the 1860s, when Henry Chadwick charged that the baseballs in use at the time were "overelastic." During those early years, the color, texture and "liveliness" of the balls varied widely according to team preference. There were many manufacturers of baseballs. Almost every city sporting a baseball team had at least one.

In 1875, the Ryan "dead-ball" was adopted as the official baseball. However, in 1876 Albert G. Spalding and his brother J. Walter Spalding of Spalding Bros., Inc., began manufacturing a ball with uniform specifications, and in 1878, the Spalding League Ball became the official baseball of the National League. The balls were *given* to the league. In fact, Spalding Bros., Inc., paid the league $1 a dozen for the privilege. The advertising paid off handsomely. When the company began operating, it boasted $800 in capital. By 1892, it had absorbed several of the largest sporting goods companies and had a capital stock worth $4 million.

In the 1950s, the tidal wave of home runs renewed speculation about the coefficient of restitution. Was the ball getting livelier? It is indeed true that home run–hitting increased consistently after the early 1900s. Figure 36 is a graph showing how the number of home runs per game per team has changed from year to year. A value of 0.2 means that teams hit an average of two home runs every 10 games, or one homer in every five games played that year. This number increased dramatically between about 1920 and 1960, reaching a value around 0.8, which translates into about 3 or 4 percent of the balls hit in fair territory being home runs. Note the steep rise in the home run rate between 1919 and 1922, the decline during World War II and the recovery after the war. Recently there has been another sharp rise.

The reason or reasons for the long-term increase in the number of home runs are difficult to pin down. Unfortunately (although Spalding Bros., Inc., claimed the manufacturing methods and materials were standard), actual quality control through regular measurements of the coefficient of restitution had not been practiced before the mid-1950s. In an article in *Collier's* magazine, Tom Meany[4] reported that Spalding Bros., Inc., claimed that there had been no changes in the specifications in manufacturing their baseballs for at least three decades. Conversely, there is a National Bureau of Standards report stating that official major league balls in 1943 were found to have coefficients of restitution of 0.41. It has been speculated that during World War II, the quality of the material used in the manufacture of baseballs was inferior to that used in other

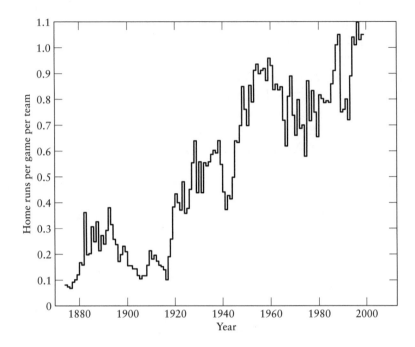

Figure 36   *The average number of home runs has increased throughout the years.*

years. This coefficient of restitution of 0.41 is much lower than the value of 0.55 that is required of the present-day ball. The fact that the home run explosion was already well on its way by the 1940s casts some doubt on the "livelier ball" theory.

The year 1987 was a banner year for home run hitting, and speculations about the increased coefficient of restitution continued to abound. Baseball manufacturers continued to deny allegations of "rabbit" balls. *USA Today* reported on July 3, 1987, that tests of 1987 and 1977 baseballs performed by Haller Testing Laboratories showed only the slightest changes. The coefficient of restitution, according to their results, *declined* by 0.4 percent during that 10-year period.

A more reasonable explanation of what is behind the trend toward more and more home runs is the "livelier ball player" theory. Babe Ruth started the trend toward more home runs in 1919. When it was pointed out to Ruth in 1930 that he made more money than President Hoover,

Ruth reportedly responded, "I had a better year." Home run hitters generally began having much better years (financially) than those who merely hit singles for higher averages. As Meany said, "The money's in the big end of the bat." When someone told former Pittsburgh Pirate first baseman and power hitter Ralph Kiner that he could raise his batting average by choking up on the bat, Kiner replied, "Cadillacs are down at the end of the bat." Hitters are concentrating more on hitting home runs. As is the case in other sports, diet and other factors have also led to healthier, stronger players.

The Best Bat Weight: From the Principles of Physics

The bat might have played at least as important a role as the ball in the home run explosion. According to statistics compiled by Hillerich and Bradsby, manufacturers of the Louisville Slugger bat, the average weight of bats used by top players decreased from about 40 ounces in the 1920s to about 32 ounces in the 1950s[4] and has kept this value to the present day. Home run production increased dramatically while the bat weight was dropping.

Baseball players, for example, Babe Ruth, have used bats as heavy as 54 ounces, but physicists have said that the optimal bat weight is only 15 ounces.[2,3] Because no one really knew what bat weight was best, over the years there has been a lot of experimenting with the bat. Most of this experimentation was illegal, because the rules say that (for professional players) the bat must be made from one solid piece of wood. To make the bat heavier, George Sisler, who was elected to the Hall Fame in 1939, pounded Victrola phonograph needles into his bat barrel and in the 1950s Ted Kluszewski of the Cincinnati Reds hammered in tenpenny nails. To make the bat lighter, many players have drilled a hole in the end of the bat and filled it with cork.[5] Detroit's Norm Cash admits to using a corked bat in 1961 when he won the American League batting title by hitting .361. However, the corked bat may have had little to do with his success, because he presumably used a corked bat the next year when he slumped to .243. Some players have been caught publicly using doctored bats. In 1987 Houston's Billy Hatcher hit the ball and his bat split open spraying cork all over the infield.

In an attempt to change the coefficient of restitution, some players have stuffed the bat with Super Balls. D. Gutman[6] says, "Back in 1974 Graig Nettles of the Yankees took a vicious swing in the fifth inning of a game against Detroit, splitting his bat as it made contact. Out bounced six Super Balls." Brancazio[3] cites an article in *Newsday* magazine saying that "a hitter sometimes hollows a bat and stuffs it with a spongy substance like cork or rubber balls to make the bat expand explosively . . . upon impact with a ball." This is nonsense, as pointed out by Brancazio. In fact, as long as the structural integrity of the bat is not destroyed by drilling too large a hole, the coefficient of restitution of the bat-ball collision is unchanged. Note that the outer surface of the bat shown in Figure 35 is undeformed during the collision with the ball. A change in the coefficient of restitution of the ball-bat combination is almost certainly not responsible for the most recent increase in home run hitting. In the remainder of this chapter and in Chapter 6 we will explore some other possibilities. Let us now see what the theory of collisions can say about this important question.

Now that we have some understanding of the coefficient of restitution, or the "bounciness" that occurs when bat meets ball, we have a way of relating the speeds of two objects before and after the collision to the physical properties of the colliding objects. This allows us to use the equation for conservation of momentum

$$m_1 v_{1b} + m_2 v_{2b} = m_1 v_{1a} + m_2 v_{2a}$$

with the equation for the coefficient of restitution

$$e = -\frac{v_{1a} - v_{2a}}{v_{1b} - v_{2b}}$$

to obtain expressions for both the bat and ball speed immediately after the collision. We find that

$$v_{1a} = \frac{(m_1 - em_2) v_{1b} + (m_2 + em_2) v_{2b}}{m_1 + m_2}$$

and

$$v_{2a} = \frac{(m_1 - em_1) v_{1b} + (m_2 + em_1) v_{2b}}{m_1 + m_2}$$

Much can be learned from these two equations, so let us dwell on them for awhile.

First, since v_{1b} is a negative number and v_{2b} is a positive number (the objects are moving in opposite directions), v_{1a}, the speed of the ball after the collision, always increases with e. Also, as long as $m_1 - em_2$ is negative (the weight of the bat is larger than $1/e$ times the weight of the ball), v_{1a} increases when $-v_{1b}$ increases. This means that a fastball can be hit harder than a change-up, all other things being equal.

Finally, the equation for v_{1a} tells us that, all other things being constant, the speed of the ball as it leaves the bat increases as the weight of the bat increases. Suppose, for example, that the bat is much lighter than the ball. Omitting the term m_2 from the immediately preceding equations, that is, assuming m_2 is close to zero, shows that in that case

$$v_{1a} = v_{1b}$$

which means the ball zooms past with undiminished speed. We also find that

$$v_{2a} = (1 - e) v_{1b} + ev_{2b}$$

(the bat flies backwards). Suppose, on the other hand, that the bat weighs very much more than the ball. In such a case, terms involving m_1 in the equations for v_{1a} and v_{2a} can be dropped. The results are then

$$v_{1a} = -ev_{1b} + (1 + e)\, v_{2b}$$

and

$$v_{2a} = v_{2b}$$

The collision with the ball does not affect the speed of the huge bat. *For a given bat speed,* this is the largest velocity that can be imparted to the ball. According to this reasoning, bats should be as heavy as the rules will allow.

There is, of course, a problem with this kind of analysis. A little thought tells us that if the bat is too heavy, the batter cannot control it well enough to make good contact with the ball. Even if he could make contact with the ball, the bat speed before contact would be smaller than that which could be attained with a lighter bat. What the analysis above does not account for is that the larger m_2, the smaller the bat speed before the collision, v_{2b}. What we have here is a case of conflicting effects.

This is as good a time as any to introduce a much-overlooked fact about the practical use of scientific theories. Scientific ideas can be used to explain what is happening in the world around us and, in many cases, to predict better ways of doing things. But we must be quite specific in the way we ask questions. Almost all interesting scientific problems involve conflicting factors. In the case of hitting a baseball with a bat, we found that, strictly from the point of view of momentum considerations, the speed of a baseball leaving the bat with a given bat speed is maximized by making the bat mass (or weight) as large as possible. However, a very large bat would be hopelessly unwieldy. From the standpoint of bat control and accuracy, the bat should be extraordinarily light. The equation

$$v_{1a} = v_{1b}$$

and the equation

$$v_{2a} = (1 - e)v_{1b} + ev_{2b}$$

as well as our intuition, tell us that such a bat would simply be knocked from a batter's hands. Surely there is an optimum bat weight between these two extremes. But optimum in what sense?

Momentum effects associated with the collision itself tell us that for a given speed, we need a massive bat. We also know, however, that the smaller the bat, the higher the bat speed that can be obtained. The speed with which a particular batter can swing his weapon depends on how much energy he can put into his swing. To resolve our conflict, and to learn more about the optimum bat and the optimum swing, we need to stop looking at only the bat and the ball. We must now consider the human being swinging the bat.

The Best Bat Weight: From the Principles of Physics and Physiology

The speed of a baseball after its collision with a bat depends on many factors, not the least of which is the weight of the bat. One professional baseball team (St. Louis Cardinals) says the weight of the bat is determined by "the player's personal preference," while another (New York Yankees) says, "Each individual player determines the style of bat he prefers." These players have very little real scientific data to help them support their preferences. In this section, we present data to help an individual player to decide if his or her preference is the most effective bat weight. Knowing the ideal bat weight can eliminate time-consuming and possibly misleading experimentation by ballplayers.

To find the best bat weight we must first reexamine the conservation of momentum equations for bat-ball collisions. For the science of baseball, the distinction between mass and weight is not critical, and so we will substitute weight for mass in the equation for the conservation of momentum to produce

$$w_1 v_{1b} + w_2 v_{2b} = w_1 v_{1a} + w_2 v_{2a}$$

Keep in mind that we are assuming the weight of the batter's arms has no effect on the collision (this may be an important assumption). We want to solve for the ball's speed after its collision with the bat, called the *batted-ball-speed*, but first we should eliminate the bat's speed after the collision, because it is not easily measured. We can use the equation for the coefficient of restitution to solve for v_{2a}, substitute the result into the equation for the conservation of momentum and solve for the ball's speed after its collision with the bat. The result is

$$v_{1a} = \frac{(w_1 - ew_2)v_{1b} + (w_2 + ew_2)v_{2b}}{w_1 + w_2}$$

This means that the ball's speed after the collision will depend on the weight of the ball and bat, the coefficient of restitution, and the precollision speeds of the ball and bat.

Paul Kirkpatrick[2] assumed that the optimal bat weight would be the one that "requires the least energy input to impart a given velocity to the ball." This definition in conjunction with the immediately preceding equation yields

$$\left[\frac{w_2}{w_1}\right]_{optimal} = \frac{v_{1a} - v_{1b}}{v_{1a} + ev_{1b}}$$

If we now make the reasonable assumptions that

$$w_1 = 5.125 \text{ oz., the weight of the baseball}$$

$$e = 0.55, \text{ the coefficient of restitution of a baseball-bat collision}$$

$$v_{1b} = -80 \text{ mph, a typical pitch speed}$$

$$v_{1a} = 110 \text{ mph, the ball speed needed for a typical home run}$$

we can solve the immediately preceding equation to find that the *optimal bat weight is* 15 ounces!

Peter J. Brancazio[3] has written an excellent theoretical analysis of bat-ball collisions. He considered not only the bat's translation but also its angular rotation about two axes. He found that the ball's speed after the collision with the bat depends on

1. the energy imparted by the body and arms;

2. the energy imparted by the wrists;

3. the speed of the pitch;

4. the point of collision of the ball with respect to

   a. the center of percussion,

   b. the center of mass,

   c. the end of the bat,

   d. the maximum energy transfer point, and

5. the weight of the bat.

However, by assuming that a professional baseball player exhibited normal values for each of these dependencies, he also concluded that the *optimal bat weight* is about 15 ounces.

These conclusions cannot help professional baseball players, who must use solid wood bats, because a 15-ounce solid wood bat would only be about 15 inches long! Such a bat would be far smaller than any bat that is now used in professional baseball. A typical major league bat weighs about 32 ounces. Babe Ruth normally used a 44-oz bat, and sometimes used bats weighing more than 50 ounces. On the other hand, a fungo bat, a bat used for hitting fly balls in practice sessions, weighs about 23 to 24 ounces. One might take this to mean it is closer to the optimum weight. However, a fungo bat is used to hit balls that have initial speeds v_{1b} of practically zero, so a reexamination of the conservation of momentum equation indicates that for that case, the bat should weigh the same as the ball, about 5.25 ounces.

These conclusions may help explain why people choke up on the bat; choking up makes the bat effectively shorter, moves the center of mass closer to the hands, thereby reducing the moment of inertia, and in essence makes the bat act like a lighter bat. Several reasons have been advanced for buying a long bat and choking up on it. First, a longer bat must be made from wood that has straighter grain. Therefore bat manufacturers use the best wood for the longer bats, and bats made from the best wood do not break as easily. Secondly, using a longer bat allows the batter to change the effective weight of the bat during his time at bat. Al Rosen recalled that Ted Williams and Mickey Mantle did not choke up with no strikes. If the pitcher got one strike on them, they choked up a half inch. If the pitcher got two strikes, they choked up an inch. This conclusion could also help explain the great popularity of aluminum bats. The manufacturers can make them lighter while maintaining the same length and width.

Both Kirkpatrick's and Brancazio's physics studies were limited by their explicit assumptions. Kirkpatrick assumed that the optimal bat was the one that required the smallest bat kinetic energy. Brancazio's calculations of the optimal bat weight were based on the assumption that the "batter generates a fixed quantity of energy in a swing," independent of the bat weight. We will now extend these studies by allowing the amount of energy imparted to the bat by the batter to depend on bat weight.

Physiologists have long known that muscle speed decreases with increasing load.[7-11] This is why bicycles have gears. The rider can keep muscle speed in its optimal range while bicycle speed varies greatly. Therefore, to discover how muscle properties of individual ballplayers affect their best bat weights, we measured the bat speeds of many batters swinging bats of various weights. We plotted the data of bat speed versus bat weight, and used this to help calculate the best bat weight for each batter.

The Bat Chooser™ Instrument[12,13]

Our instrument for measuring bat speed, the Bat Chooser,* has two vertical light beams, each with associated light detectors (similar to the electric eyes on elevator doors). The subjects were positioned so that when they swung the bats the center of mass of each bat passed through the light beams. A computer recorded the time between interruptions of the light beams. Knowing the distance between the light beams and the time required for the bat to travel that distance, the computer calculated the speed of the bat's center of mass for each swing. Our computer sampled every 16 microseconds. In all cases our velocities are accurate to better than 1 percent.

Note: This book is a scientific analysis of baseball. To make the mathematics easier to follow, in the early part of this chapter, we make our calculations for a ball that hits the bat at its center of mass. Consequently, all of the bat speeds given in this chapter are speeds for the center of mass of the bat. Many *experimental* studies of baseball give bat speeds for the *sweet spot* of the bat. (There are many definitions for the sweet spot of the

*Bat Chooser is a trademark of Bahill Intelligent Computer Systems.

bat.[14]) To compare our speeds for the center of mass to experimental studies giving the speed of the sweet spot, multiply our numbers by about 15 percent. (Because the swing of the bat is a combination of translation and rotation, a combination that varies from person to person and from bat to bat, it is impossible to analytically specify this correlation factor. Our 15 percent figure is based on a melding of theoretical analyses and experimental measurements. In our experiments with 340 swings by 11 people the difference was 12 percent with a standard deviation of 6 percent.) For example, for a bat swing where we report a center of mass bat speed of 50 mph, the speed of the sweet spot would be about 58 mph.

To hit a home run in most major league parks it takes a center of mass bat speed of about 50 mph, which combined with a pitch speed of about 90 mph would produce an initial batted-ball speed of 100 mph, which would put the ball 15 feet high, 330 feet from the plate.

As an aside, in 1999, television programs started reporting "bat speeds." These numbers must be viewed with suspicion. They often reported consecutive swings with speeds of, for example, 74, 84 and 94 mph. Professional baseball players do not have such variability in their swing speeds. Consecutive swings typically vary by only a few mph. Batters are trained to make their swings repetitive, that is with low variability. When the speed of the pitch varies from 70 to 80 to 90 mph, the batters adjust the *time of onset* of their swing, not the speed of the swing. These television reports are probably measuring the speed of the sweet spot on one swing, the speed of the tip of the bat on the next swing and the speed of the reflection off the tip of the bat on the next. But they do not tell us what they are measuring, because they do not know what they are measuring. When the television reports get into the range of 50 to 60 mph, or when they tell us what they are actually measuring, then we can start to believe them.

In our experiments, each player was positioned so that bat speed was measured at the point where the subject's front foot hit the ground. This is the place where most players reach maximum bat speed and therefore where they hit the ball with maximum force.

We told the batters to swing each bat as fast as possible while still maintaining control. We told the professionals to "Pretend you are trying to hit a Randy Johnson fastball."

In our experiments, each adult subject swung six bats through the light beams. The bats varied from superlight to superheavy, yet they had similar lengths and weight distributions. In our experiments we used about six dozen different bats. We used aluminum bats, wood bats, plastic bats, infield fungo bats, outfield fungo bats, bats with holes in them, bats with lead in them, major league bats, college bats, softball bats, Little League bats, brand new bats and bats more than 40 years old. In one set of experiments, we used the six bats shown in Figure 37 and described in Table 2. These bats were about 35 inches long, with the center of mass about 23 inches from the end of the handle.

For Little League players we changed to a different set of bats. They had to be lighter and fewer in number. For our final experiments, we

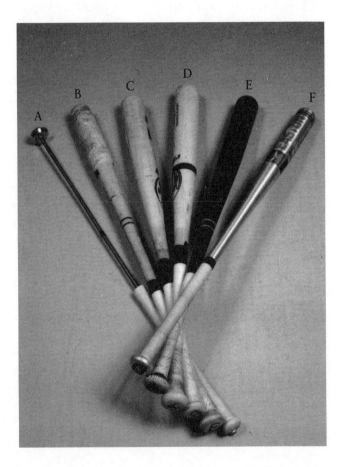

Figure 37   The bats used in some experiments; the lightest is on the left, the heaviest on the right. These bats are described in Table 2. (Photo credit Richard Harding.)

Table 2   Characteristics of the Six Adult's Bats

Name	Weight (oz)	Length (in)	Distance From the End of the Handle to Center of Mass (in)	Composition
F	49.0	35.0	22.5	Aluminum bat filled with water
E	42.8	34.5	24.7	Wood bat drilled and filled with lead
D	33.0	35.5	23.6	Regular wood bat
C	30.6	34.5	23.3	Regular wood bat
B	25.1	36.0	23.6	Wood fungo bat
A	17.9	35.7	21.7	A wooden bat handle mounted on a threaded steel lamp pipe with a 6-oz weight attached to the end

used the set described in Table 3. However, even with this set we saw signs of fatigue in half our subjects.

In a 20-minute interval of time, each subject swung each bat through the instrument five times. The order of presentation was randomized. The selected bat was announced by a speech synthesizer; for example, "Please swing bat Hank **A**aron, that is bat **A**." We recorded the bat weight and the linear velocity of the center of mass for each swing.

The Force-Velocity Relationship of Physiology

When bat speeds measured with this instrument were plotted as a function of bat weight, we obtained the typical muscle force-velocity relationship shown in Figure 38. The ball-speed curve and the term *ideal bat weight* shown in this figure will be discussed in a later section. This force-velocity relationship shows that the kinetic energy ($\frac{1}{2} mv^2$) put into a swing was zero when the bat weight was zero, and also when the bat was so heavy that the speed was reduced to zero. The bat weight that allowed the batter to put the most energy into the swing, the *maximum-kinetic-energy bat weight*, occurred somewhere in between. This led to the suggestion that the batter might choose a bat that would allow maximum kinetic energy to be put into the swing. Figure 39 shows the kinetic energy of the bat as a function of bat weight for a member of the San Francisco Giants. This batter could impart the maximum energy to a bat weighing 46.5 ounces.

This maximum-kinetic-energy bat weight does not, however, tell us the bat weight that will make the ball leave the bat with the highest speed. To calculate this weight we must couple the muscle force-velocity relationship to the equations for conservation of momentum. We can then solve the resulting equations to find the bat weight that would allow a batter to produce the greatest batted-ball speed. This would, of course, make a potential home run go the farthest, and give a ground ball the maximum likelihood of getting through the infield. We call this weight the *maximum-batted-ball-speed bat weight*.

Coupling Physics to Physiology

Next, we coupled the equations of physics to the equations of physiology. Over the last half century physiologists have used three equations to describe the force-velocity relationship of muscles: that for the straight line ($y = Ax + B$), that for the hyperbola (($x + A$) ($y + B$) = C) and that for the exponential ($y = Ae^{-Bx} + C$).[11] Each of these equations has been best for some experimenters, under some conditions, with certain muscles, but usually the one for the hyperbola fits the data best. In our exper-

Name	Weight (oz)	Length (in)	Distance to Center of Mass (in)	Composition
A	40.2	29.9	17.8	Wood bat with iron collar
B	5.2	31.3	17.6	Plastic bat
C	25.1	28.0	17.3	Wood bat
D	21.1	28.8	17.0	Aluminum bat

Table 3   Characteristics of Bats Used by Little Leaguers

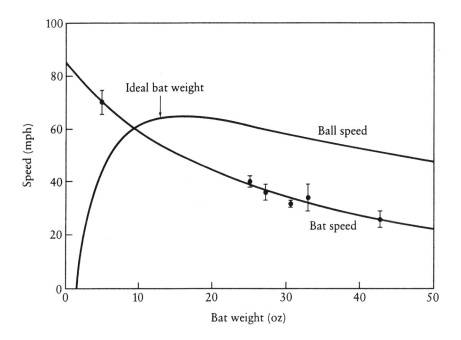

Figure 38   *Bat speed and calculated ball speed after the collision (called the batted-ball speed) both as functions of bat weight for a 40 mph pitch to Alex, a 10-year-old Little League player. The dots represent the average of the five swings of each bat; the vertical bars on each dot represent the standard deviations. These data were collected with a different set of bats than that described in Table 3. (From A. T. Bahill and W. J. Karnavas, Biological Cybernetics, 62, 1989, 89–97.)*

iments we fit all three and chose the equation that gave the best fit to the data of each subject's 30 swings. For the data of Figure 39, the best fit was: bat speed (in mph) = -0.34 bat weight (in oz) + 48, or

$$v_{2b} = -0.34w_2 + 48$$

Next, we substituted this relationship into the equation

$$v_{1a} = \frac{(w_1 - ew_2)v_{1b} + (w_2 + ew_2)v_{2b}}{w_1 + w_2}$$

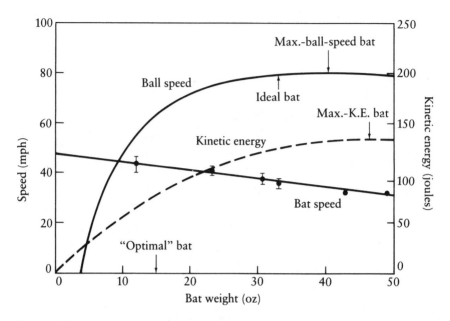

Figure 39 *Bat speed, kinetic energy given to the bat and calculated batted-ball speed, all as functions of bat weight for a 90 mph pitch to a member of the San Francisco Giants baseball team. Data for other professional baseball players were similar. These data were collected with a different set of bats than that described in Table 2. (From A. T. Bahill and W. J. Karnavas, Biological Cybernetics, 62, 1989, 89–97.)*

yielding

$$v_{1a} = \frac{(w_1 - ew_2)v_{1b} + (w_2 + ew_2)(Aw_2 + B)}{w_1 + w_2}$$

We then took the derivative with respect to the bat weight, set this equal to zero, and solved for the maximum-batted-ball-speed bat weight. The result is

$$w_{2mbbs} = \frac{-w_1A - \sqrt{w_1^2A^2 - Aw_1(B - v_{1b})}}{A}$$

For the data of Figure 39, this was 40.5 oz.

The physics of bat-ball collision predicts an optimal bat weight of 15 oz. For the professional baseball player of Figure 39, the physiology of the muscle force-velocity relationship reveals a maximum-kinetic-energy bat weight of 46.5 oz. When we coupled the equation $v_{2b} = -0.34w_2 + 48$, fit to the force-velocity data of Figure 39 to the equation derived from the coefficient of restitution and the principle of conservation of momentum for bat-ball collisions, we were able to plot the ball speed after the collision as a function of bat weight, also shown in Figure 39. This curve shows that the maximum-batted-ball-speed bat weight for this subject was 40.5 oz. which is heavier than that used by most batters. However, this ball speed curve is almost flat between 34 and 49 oz. There is only a 1.3 percent difference in the batted-ball speed between a 40.5-oz bat, and the 32-oz bat normally used by this player. Evidently the greater control permitted by the 32-oz bat outweighs the 1.3 percent increase in speed that could be achieved with the 40.5-oz bat.

Ideal Bat Weight™

The maximum-batted-ball-speed bat weight is probably not the best bat weight for any player. A lighter bat will give a player better control and more accuracy. Obviously, a trade-off must be made between maximum batted-ball speed and controllability. Because the batted-ball speed curve of Figure 39 is so flat around the point of the maximum-batted-ball-speed bat weight, we believe there is little advantage in using a bat as heavy as the maximum-batted-ball-speed bat weight. Therefore, we have defined the *Ideal Bat Weight** to be the weight at which the ball speed curve drops 1 percent below the speed of the maximum-batted-ball-speed bat weight. Using this criterion, the ideal bat weight for this subject is 33 ounces. We believe this gives a reasonable trade-off between distance and accuracy. Of course, this is subjective and each player might want to weigh the two factors differently. It does, however, give a quantitative basis for comparison. The player of Figure 39 was typical of the *San Francisco Giants* players whom we studied, as shown in Table 4, except that his swings were slower but more consistent than most. He is a control hitter.

For contrast, in Figure 40 we show the data of a San Francisco slugger who was less consistent. However, his slugging average was over

*Ideal Bat Weight is a trademark of Bahill Intelligent Computer Systems.

Table 4	Summary of Data for the 28 San Francisco Giants						
Maximum Batted-ball Speed (mph)		Ideal Bat Weight (oz)		Actual Bat Weight (oz)		Maximum Kinetic Energy (joules)	
Average	Range	Average	Range	Average	Range	Average	Range
99	80–122	31.7	26.25–37.00	32.3	31–34	270	133–408

.600. His data are best fit with bat speed (in mph) = −0.39 bat weight (in ounces) + 63. Substituting this into the equation for v_{1a} and performing our other calculations yield an ideal bat weight of 32 ounces. It is surprising that although the two players and their data as shown in Figures 39 and 40 are so dissimilar, their ideal bat weights are nearly the same. These data contrast dramatically with data from the 10-year-old boy of Figure 38. His data are best fit with this hyperbola:

$$(w_{bat} + 28.0) \times (\text{bat speed} + 12.8) = 2728$$

Substituting this into the equation for v_{1a} and performing our other calculations yields an ideal bat weight of 15 ounces.

The ideal bat weight varies from person to person. Table 5 shows the means and standard deviations of ideal bat weights for batters in various organized leagues. These calculations were made with the pitch speed each player was most likely to encounter, for example, 40 mph for Little League and 20 mph for university professors playing slow pitch softball.* Ideal bat weight is specific for each individual, but it does not appear to be correlated with height, weight, age, circumference of the upper arm or any combination of these factors, nor is it correlated with any other obvious physical factors.

To further emphasize the specificity of the ideal bat weight calculations, we must display individual statistics rather than averages and standard deviations. In Figure 41, we compare the ideal bat weight with the weight of the actual bat used by the players before our exper-

*The coefficient of restitution of a softball is smaller than that of a baseball, but this did not affect our calculations, because ideal bat weight is almost independent of the value of the coefficient of restitution.

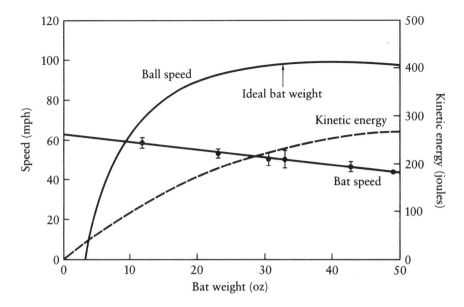

Figure 40   Bat speed, kinetic energy and calculated batted-ball speed for another member of the San Francisco Giants baseball team.

iments. This figure shows that most of the players on the San Francisco Giants baseball team are using bats in their correct range. The dashed lines in this figure (derived from data and calculations not shown in this chapter) delineate the range of bat weights that we recommended to their management. We recommended that batters above the upper dashed line switch to heavier bats and that batters below the lower dashed line switch to lighter bats.

Not only is the ideal bat weight specific for each player, but it also depends on whether the player is swinging right- or left-handed. We measured two switch-hitters (one professional and one university ballplayer who later had a long professional career). One player's ideal bats weights were 1 ounce different and the other's were 5 ounces different. Switch-hitters were so different when hitting right- and left-handed that we treated them as different players.

Extrapolating from the equation

$$w_{2mbbs} = \frac{-w_1 A - \sqrt{w_1^2 A^2 - A w_1 (B - v_{1b})}}{A}$$

Table 5 Ideal Bat Weights

League	Mean Ideal Bat Weight (oz)	Standard Deviation	Typical Pitch Speed (mph)	Number of Subjects
Professional, major league	31.1	3.6	90	27
University baseball	29.4	4.2	80	9
University softball	29.7	4.3	60	19
Junior League, ages 13–15	21.7	4.9	55	6
Little League, ages 11–12	21.3	2.9	50	34
Little League, pages 9–10	21.5	3.7	40	29
Little League, ages 7–8	19.0	3.1	35	27
Slow Pitch Softball	25.7	3.7	20	12

shows that the ideal bat weight also depends on pitch speed. Figure 42 shows this dependence of ideal bat weight on pitch speed for the ballplayer of Figure 39. This figure also shows the resulting batted-ball speed after a collision with a bat of the ideal weight. Such curves were typical of all our subjects.

This figure shows that the ideal bat weight increases with increasing pitch speed. This means that even if they could swing 33-oz bats, Little Leaguers should use lighter bats, because the pitch speeds are lower. However, when this figure is used to identify the ideal bat weight for a particular individual, the results may seem counterintuitive. When the opposing pitcher is a real fireballer, the coach often says, "Choke up [i.e., get a lighter bat] so you can get around on it." In such situations, the coach is changing the subjective

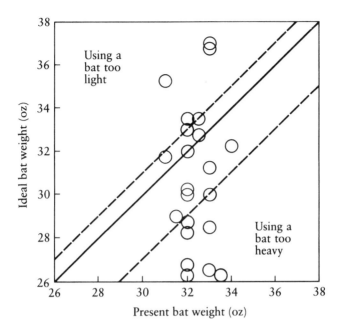

Figure 41   *Ideal bat weight versus actual bat weight for the San Francisco Giants. Most of them are using bats in their recommended range. (From A. T. Bahill and W. J. Karnavas,* Biological Cybernetics, *62, 1989, pp. 89–97.)*

weighting of bat control versus distance. He is asking the player to drop his criterion to 2 or 3 percent below maximum-batted-ball-speed bat weight so he can get better bat control. Since the batted-ball speed depends on both the pitch speed and the bat weight, the batter can afford to choke up when facing a fast pitcher, knowing that the ball will go just as far as it would if he were not to choke up when facing a slower pitcher.

One more result that came from our studies is that proficient batters have consistent swings. They are machinelike. They train this machinelike precision. Compare the height of the data crosses (the standard deviation) of the swings of the Little League player of Figure 38, who had been playing ball for five years, with the height of the crosses for the swings of the professional player who had been playing ball for 24 years (Figure 39). Consistency is important and the professional player shows this consistency.

In 1988, we measured the bat swings of the women on the University of Arizona softball team. For most players we recommended bats between 24 and 30 ounces. In almost all cases, the recommended bat

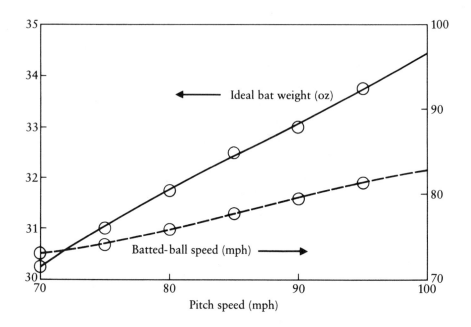

Figure 42   Ideal bat weight and batted-ball speed both as a function of pitch speed for the professional baseball player of Figure 39. (From A. T. Bahill and W. J. Karnavas, Biological Cybernetics, 62, 1989, 89–97.)

was significantly lighter than the one they were using. The players replied, "But what are we to do? The lightest bat on the rack is 30 ounces." We responded, "Our recommendations for tomorrow are not tied to yesterday's technology." In 1994, we again measured the women on the University of Arizona softball team. Again we recommended bats in the range of 24 to 31 ounces, but this time most of them were already using bats within one ounce of what we recommended.

Generalizations and Limitations

They won the collegiate World Series in 1991, 1993, 1994, 1996 and 1997. In 1999 we measured them again. We recommended bats in the range of 25 to 29 ounces. But this time most of them were using bats in the range of 23 to 25 ounces. The bat manufactures have finally caught up and indeed surpassed our recommendations for lighter weight bats. It is not surprising that in a game that is more than 100 years old, played by players being paid millions of dollars per year, that professional athletes, without the benefit of scientists and engineers, have found that their best bat weights are between 31 and 33 ounces. However, it is interesting to note that, given the relative newness of the aluminum bat and the fact that the players using aluminum bats are amateurs, Little League and slow pitch softball players are just beginning to get bats that are light enough for them. The lightest Little League approved bat that we have seen is 15 ounces. The lightest legal softball bat that we have seen is 23 ounces. (However, these numbers are decreasing at the amazingly high rate of about an ounce per year.)

For the Little Leaguer of Figure 38, the batted-ball speed varies greatly with bat weight. This means that it is very important for him to have the correct weight bat. However, for most professional baseball players, once the bat is in the correct range, the batted-ball speed varies little with bat weight, as shown in Figure 39. That player could use any bat in the range 33 to 40 ounces and there would be less than a 1 percent change in batted-ball speed. This fact is not in the literature, and it could not be determined accurately by experimentation by ballplayers on the ball field. For example, imagine an experiment where a pitcher alternately throws 20 white balls and 20 yellow balls to a batter who alternately hits with a 32- and a 34-oz bat. Imagine then going into the outfield and looking at the distribution of distances of the balls. You would not see the yellow balls or the white balls consistently farther out. Variability in the pitch and the location of the contact point between the bat and the ball would obscure any differences. However, with our Bat Chooser instrument we can accurately measure bat speed and calculate the resulting batted-ball speed. Our calculations show that this curve is flat. This knowledge should help professional batters eliminate futile experimentation with bat weights, while attempting to get higher batted-ball speeds. As long as the player uses a bat in the flat part of his curve, there will be less than a 1 percent variation in batted-ball speed caused by varying bat weights. What does a 1 percent decrease in batted-ball speed mean? A ball that would normally travel 333 feet would only travel 330 feet. This is not terribly important.

In our studies, we measured bat speed as a function of bat weight. Next, we coupled these measurements to the equations of physics and physiology to determine the ideal bat weight for each individual batter. We can say nothing about the "feel" of a bat; this is a psychological variable that we cannot measure. We cannot say anything about the grain or strength of the bat. In this study, we say little about choking up on the bat. We treat a 33-oz bat choked up 1 inch as a 32-oz bat. We are not concerned with the availability of bats. Our recommendations are independent of what equipment is actually available. We have tried to make sure that our solutions to tomorrow's problems are not stated in terms of the hardware that was available yesterday. We have no means of assessing the accuracy of the swing. Throughout our analysis, we assume that it is easier to control a lighter bat than a heavier one.

The experiments discussed in this section measured the linear velocity of the center of mass of the bats. It is obvious that in addition to this translation the bat also rotates about two different axes. However, our results derived from only linear velocity agree in most details with those Brancazio derived using angular velocity. The exception is that for a rotating bat, the place where the ball hits the bat becomes important, as will be shown. If the ball hits the bat at its center of mass, the results of the linear approximation are the same as those derived treating translation and rotation separately. Furthermore, in later experiments where we measured the speed of the sweet spot of the bat and the speed of the center of mass, we found little differences in the ideal bat weight.

We have also neglected the effect of air resistance on pitch speed. We calculated the ideal bat weights of the major league players based on a pitch speed of 90 mph. If the ball were going 90 mph when it hit the bat, it would indeed be a fast pitch, because the ball loses about 10 percent of its speed on its way to the plate. If we decreased our 90 mph figure by 10 percent, the ideal bat weights would decrease by approximately 1 ounce.

Our data have low variability for physiological data. For the data of Figure 39, the standard deviations are about ±5 percent. However, the repeatability of our experiments is not as good. On any given day the data are repeatable, but tests run 1, 2 or 12 months apart differ by as much as 20 percent in bat speed for any given bat. However, in spite of these large differences in bat speed, the calculated ideal bat weight varies by only an ounce or two. We are still looking for the sources causing the lack of repeatability. We think the most likely causes are warm-up status, adrenaline, positioning of the subject and fatigue.

Our experiments were done indoors. Some of the ballplayers thought things would be different out on the field swinging against a real pitcher. To study this effect we took the equipment out to the ball field. Immediately after an intrasquad game, we measured the bat speeds of four players while they hit the ball. For each player these data fell within the range of his data collected in the laboratory six months earlier and one week later.

Two Types of Force-Velocity Curves

We found two classes of batters: those whose force-velocity relationships were fit best with straight lines (like Figure 39), and those that were fit best with hyperbolic curves (like Figure 38). The data of the Little Leaguer shown in Figure 38 are fit best with this hyperbola: $(w_{bat} + 28.0)$ × (speed + 12.8) = 2728. The data of his brother, also a Little Leaguer, collected on the same day are fit best with a straight line. Most Little Leaguers were fit best with hyperbolas, half of our college players were fit best with hyperbolas and one fourth of our major leaguers were fit best with hyperbolas.

The data for two more major leaguers are shown in Figure 43. The top figure is for a young Matt Williams (a quick shortstop who later in his career turned into a power-hitting third baseman): the hyperbolic fit (solid line) is 35 percent better (in a mean squared error sense) than the straight line fit (dotted line). The bottom figure is for Candy Maldonado (a nonquick power hitter); his data are fit best with a straight line. Both of these players have maximum batted-ball speeds above 110 mph (in Chapter 6 we show that 110 mph is sufficient for a home run in most ballparks). And indeed both of these players have hit more than 20 home runs in a season.

The two professional players of Figure 43 both normally used 33-oz bats. As shown in this figure, they swung our 31- and 33-oz bats at the same speed. Within the limits of statistical variability, we can say that for bats in the range that they normally use, the two players swing with the same speed. However, our recommendations for these two players were quite different. We recommended that the player of the bottom figure keep his 33-oz bat, but that the player of the top figure change to a 31-oz bat. Why did this happen? In our experiments we gave them very heavy and very light bats, unlike any bats they have ever encountered. We collected data when they swung these bats and we made mathematical models of each human, for example, the solid lines of Figure 43. From these models, we made predictions about each individual's ideal bat weight. Modeling is a common scientific technique. Scientists and engineers build models and use the models to make predictions about the physical world. That is what we have

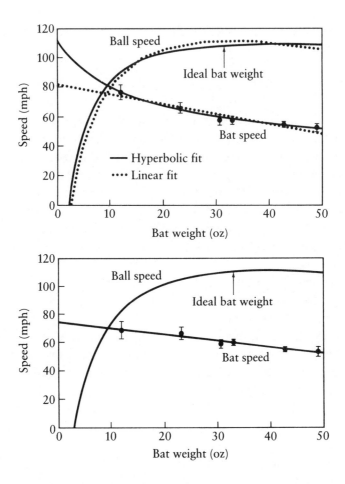

Figure 43   Bat speed and calculated batted-ball speed as functions of bat weight for a 90 mph pitch to two major league hitters. The player of the top graph was "quick," while the player of the bottom graph was not. (From A. T. Bahill and W. J. Karnavas, Biological Cybernetics, 62, 1989, pp. 89–97.)

done here. We have made models of the individual players and have used these models to predict their ideal bat weights. Such predictions could not be made using just experimental data from bats that they normally use.

We tried to correlate the type of curve that gave the best fit (i.e., straight line or hyperbola) with height, weight, body density, arm circumference, present bat weight, batting average, slugging average, running speed, and so on, but had no success. However, we noted that the subjects who were fit best with hyperbolas were described by their coaches as being "quick." Quickness is not the same as running speed, but it is related. Quickness is easy to identify but hard to define. Shortstops, centerfielders and leadoff batters are usually quick. Coaches easily identified their quick players, but when asked to explain why they called these players *quick*, they waffled. Their uneasy verbalizations include phrases like "they react quickly," "they move fast," "they steal many bases," "they get into position to field the ball quickly," "they swing the bat fast" and "they beat out bunts." But all these phrases describe resulting behavior, not physiological characteristics. So we decided to measure eye-hand reaction time and try to correlate it with the bat swing data. Eye-hand reaction time was measured by:

1. Holding a meter stick in front of a subject. Instructing him or her to place the opened index finger and thumb at the 50-cm mark and to watch the fingers of the experimenter, who holds the end of the meter stick.

2. Having the subject close his fingers as quickly as possible when the experimenter releases the meter stick and it begins to fall.

3. Noting the place where the subject catches the meter stick. The place indicates eye-hand reaction time ($d = \frac{1}{2}at^2$).

4. Giving each subject two warm-up trials, and then collecting data for 11 trials.

5. Selecting the median value of these 10 trials as our measure of quickness.

The median eye-hand reaction time for the quick boy of Figure 38 was 143 milliseconds (ms); for his nonquick brother it was 256 ms. We collected eye-hand reaction times for 21 of the San Francisco Giants and compared it with the percentage superiority of the hyperbolic fit over the straight line fit. We found that the physiologic parameter that best differentiates between players whose data can be fit best with a straight line and those who require a hyperbola is the eye-hand reaction time.

For our nonquick subjects the weight of the bat seemed to have little effect on how they swung it. They swung all bats with about the same speed, and their data were fit best with a straight line, as shown in the bottom graph of Figure 43. For our quick subjects the weight of the bat was a limiting factor; speed depended on weight. The curves had steep slopes and needed hyperbolas to fit the data, as shown in the top graph of Figure 43. We hypothesize that the quick people change their control strategies when given different bats; whereas nonquick people do not change their strategies—they swing all bats the same way.

Resume

The physics of bat-ball collisions without consideration of physiology (specifically the equations for conservation of momentum and the coefficient of restitution) predicted an *optimal bat weight* of 15 ounces. The physiology of the muscle force-velocity relationship showed that the professional baseball player of Figure 39 could put the most energy into a swing with a 46-oz bat; that is, the *maximum-kinetic-energy bat weight* was 46 ounces. Coupling physics to physiology showed his maximum-batted-ball-speed bat weight to be 40 ounces. Finally, trade-offs between maximum ball speed and controllability showed that his *ideal bat weight* was 33 ounces. These experiments explain why most adult batters use bats in the 28- to 34-oz range, they explain the variability in human choice of bat weight and they suggest that there is an ideal bat weight for each person.

A Database Analysis as an Alternative to the Bat Chooser

Hitting a baseball is the hardest act in all of sports. This act is easier if the right bat is used, but it is difficult to determine the best bat for any individual. Therefore we developed the Bat Chooser to measure the swings of an individual, make a model for that person and recommend a specific bat weight for that person.[12,13] However, this system is not conveniently available to most people. So we used our database of 200 people who had been measured with our system and created simple equations that can be used to recommend a bat for an individual using common parameters such as age, height and weight. This section on the database analysis is based on A. Terry Bahill and M. Morna Freitas[14] (1995).

A database approach was used to analyze the interrelations of our data. A database of 36 factors on 200 subjects was constructed. This database was iteratively simplified and finally the following *Best-Fit Model* was proposed for recommending a bat weight.

Recommended Bat Weight = A (age) + B (weight) + C (height) + D (typical pitch speed) + E (reaction time)

where Recommended Bat Weight is in ounces, age is in years (an age of 26.5 was used for all subjects over 26, because it produced a fit with the minimum error), body weight is in pounds, height is in inches, typical pitch speed comes from Table 5, and reaction time is in milliseconds.

The Best-Fit Model gave a good fit to the data, but it is complex because it uses reaction time as a parameter, which restricts its usefulness because most people do not have such data. Therefore, we eliminated reaction time.

Next, we sought further ways to simplify the model. The *dependence* of a parameter was given by SigmaPlot™, a scientific statistical software package, and it indicated the dependencies of parameters on one another: this technique suggested that weight be eliminated. *Influence coefficients* were generated by ModelWare™ and they quantify the degree of influence exerted among system variables: this technique also suggested that weight be eliminated. Finally, using NeuralWare™, a neural network was trained using these five columns as inputs to predict the ideal bat weight output column. Then one input column was removed at a time and we noted which columns affected the error the most. This technique also suggested that weight be eliminated. Therefore, we eliminated weight. This reduction in parameters produced the *Parsimonious Model*.

Recommended Bat Weight = A (age) + C (height) + D (typical pitch speed)

The Parsimonious Model fits the data almost as well as the Best-Fit Model, but it is not as complex because it uses fewer parameters, and only those readily available to unsophisticated users. Finally, we again used dependencies, influence coefficients and the neural net to eliminate age and produce the *Simple Model* that can be used to recommend a bat in a quick and easy fashion.

Recommended Bat Weight = C (height) + D (typical pitch speed)

When a multiple linear regression analysis of the full databases was performed, these models resulted.

Best-Fit Model:

Recommended Bat Weight = 0.1163 (age) + 0.0696 (weight) + 0.0108 (height) + 0.1106 (typical pitch speed) + 0.0468 (reaction time)

We think that these recommendations are accurate to within 2 ounces, which is certainly superior to the present technique, which is "Use the same bat as the kid down the street who hits home runs." For completeness, we now present the two simpler models derived by a multiple regression analysis on the 200-subject database.

Parsimonious Model:

Recommended Bat Weight = 0.2588 (age) + 0.2592 (height) + 0.0996 (typical pitch speed)

Simple Model:

Recommended Bat Weight = 0.3070 (height) + 0.1215 (typical pitch speed)

These models should be useful: a bat can be suggested for an individual using only a simple calculator.

However, we wanted to make the models even simpler. Therefore, the 200 subjects in the database were divided into eight groups. Then we took the Best-Fit Model, restricted the number of parameters to two, restricted the parameters to be integers and then found the values that produced the least mean-squared error between the model and the data. Thus, we created eight *Integer Models*, or rules of thumb, for use in giving quick advice to a person choosing a bat: Table 6 shows these extremely simple models.

Ten years ago, no bats were available that fit these recommendations. Five years ago, few bats were available that fit these recommendations. Now, there are bats that fit all these recommendations. Perhaps, the most useful entry in Table 6 shows that for a typical nine- or ten-year-old

Recommended Bat Weight = Height/3 + 4

Table 6 Simple Integer Models for Recommending Bat Weights	
Group	**Recommended Bat Weight (oz)**
Baseball, Major League	Height/3 + 7
Baseball, Amateur	Height/3 + 6
Softball, Fast Pitch	Height/7 + 20
Junior League (13–15 years)	Height/3 + 1
Little League (11–12 years)	Weight/18 + 16
Little League (9–10 years)	Height/3 + 4
Little League (7–8 years)	Age*2 + 4
Softball, Slow Pitch	Weight/115 + 24

Recommended Bat Weight (ounces), age (years), height (inches) and body weight (pounds).

Clearly, this is a model that many sporting goods store employees and coaches should find very useful when asked to recommend a bat for a child.

Rotational Motion of the Bat

This section is technical and may be skipped without losing continuity. The equation for acceleration,

$$F = m\frac{v - v_0}{t}$$

the equation for conservation of momentum

$$m_1 v_{1b} + m_2 v_{2b} = m_1 v_{1a} + m_2 v_{2a}$$

and the intervening equations at the beginning of this chapter are true as long as the velocities in those equations are the velocities of the centers of mass of the two bodies. The momenta in this case are called *linear* or *translational* momenta. The sum of the translational momenta of two colliding objects is conserved during a collision. However, the swing of a baseball bat exhibits two types of motion: translational and rotational. This is illustrated in Figure 44. To move the bat from position A to position B. one can first *rotate* the bat about the center of mass and then *translate* the center of mass. Each type of motion has associated with it a type of kinetic energy. The kinetic energy of translation is

$$KT = \frac{m_2 v_{2c}^2}{2}$$

where m_2 is the mass of the bat and v_{2c} is the speed of the center of mass of the bat. The kinetic energy of rotation is

$$KR = \frac{I_c \omega_c^2}{2}$$

where ω_c is the rotation rate about the center of mass and I_c is the moment of inertia. The moment of inertia is related to the mass of the bat and its shape. In swinging a baseball bat KR is usually much smaller than KT.

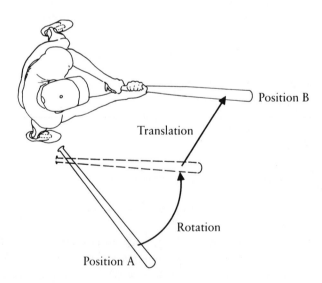

Figure 44   *The batter imparts two types of kinetic energy to the bat— rotational and translational.*

As we have seen, another theorem from physics, the work-kinetic energy theorem, tells us that when work is done on a body, kinetic energy is produced in an amount equal to the work done:

$$Work = KT + KR$$

Clearly, there is some maximum amount of kinetic energy that a given player can put into the swing of a bat. Furthermore, one can imagine that the more effort put into maximizing the kinetic energy of the bat, the less bat control the hitter has.

In order to understand the importance of this rotational motion, we must pause for a moment and consider how rotational and translational motion are related and how forces create rotational motion.

First, consider the general case shown in Figure 45 in which a rigid body of arbitrary shape is supported at a point A, which we take to be its center of rotation. Suppose a force F is applied at a point B at a distance H from A and in a direction perpendicular to a line joining A and B. The product HF is called the *torque T*. We are interested in how the torque produces rotational motion about the point A.

First, note that if the bar rotates about the point A the speeds of different points on the body will be different. In fact, a little thought reveals that a point, P_1, that is twice as far from A as another point, P_2, will move twice as far along a circular arc as will point P_2 in a given time. Therefore, P_1 must move twice as fast as P_2. In general,

$$\frac{\text{Speed of point } P_1}{\text{Speed of point } P_2} = \frac{\text{distance from } A \text{ to } P_1}{\text{distance from } A \text{ to } P_2}$$

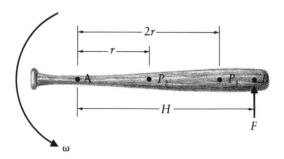

Figure 45   *When a force is applied at point B and the bat is pivoted at point A, the bat will acquire angular velocity,* ω.

The *angular velocity* ω is defined as

$$\omega = \frac{v}{D}$$

where v is the translational speed at any point on the body and D is the distance from A to that point. By comparing the two immediately preceding equations, we can see that the angular velocity is the same at all points on the rigid body. The angular velocity must be given in dimensions of radians per second, where one rotation per second is equal to 2π radians per second.

When a torque is applied to a body, the equation for acceleration

$$F = \frac{(v - v_0)m}{t}$$

cannot be applied directly because each point on the body is moving at a different speed. However, since the rotational velocity is the same at all points on the body, we can take a different approach. We view the body as being made up of a large number of small masses, Δm, each located at a distance r from the center of mass. The momentum of each piece is then $\Delta m \times r \times \omega$. And the angular momentum of each piece is defined as $r \times \Delta m \times r \times \omega$, or $\Delta m \times r^2 \times \omega$ and the sum of the angular momenta of all the pieces is called the *angular momentum* of the body:

$$AM = \Sigma \, (r^2 \, \Delta m \times \omega)$$

The symbol Σ means "sum of." Since ω is the same for each piece,

$$AM = \omega \times \Sigma \, r^2 \, \Delta m$$

Just as a force operating over time creates momentum, a torque T creates angular momentum. This is described mathematically by an equation similar to the equation for acceleration:

$$Tt = (\omega - \omega_0) \, \Sigma r^2 \Delta m$$

If the initial angular momentum is $\omega_0 \Sigma r^2 \Delta m$, the angular momentum after time t, when the torque T is applied, is $\omega \Sigma r^2 \Delta m$.

The quantity $\Sigma r^2 \Delta m$ is clearly a function of both the mass of the object and the distribution of the mass. It will require more torque to speed up (in rotational motion) an object having most of its weight far from its center of mass than an object with most of its weight close to its mass center because $\Sigma r^2 \Delta m$ will be larger in the former than in the latter. The quantity $\Sigma r^2 \Delta m$ is given a special name; it is called the *moment of inertia* about the center of mass or I_0.

When a real baseball player swings a real bat, the motion of the bat consists of a combination of two rotational motions, one centered in the batter's body and the other in the batter's wrists. We are now in a position to reconsider the baseball-bat collision accounting for these rotational motions. The following discussion follows the analyses of Peter J. Brancazio[3,15] and H. Brody.[16] Referring to Figure 46, we designate the distance from the center of rotation of the batter's body to his wrists as R, and we consider this distance to be constant, although it surely varies somewhat during the batter's swing. Similarly, we designate the distance from the wrists to the center of mass of the bat as H and the distance from the center of mass of the bat to the point where the ball hits the bat as B. The angular velocity of the batter's body and arms is ω_{body} so the linear velocity of his hands at a distance R from his body is

$$v_{hands} = R\,\omega_{body}$$

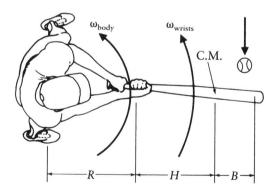

Figure 46   A batter swinging a bat will produce rotations of the bat with both his body and his wrists. C.M. designates the center of mass of the bat.

The velocity of the bat's center of mass due to the rotation of his body and arms is $(R + H)\,\omega_{body}$ while the velocity of the center of mass due to wrist rotation is $H\,\omega_{wrists}$. The sum of these is the velocity of the center of mass of the bat. Using the symbol v_{2b} for the velocity of the center of mass of the bat before collision, as we did earlier in this chapter,

$$v_{2b} = (R + H)\,\omega_{body} + H\omega_{wrists}$$

Following the same reasoning, the velocity of the point B due to ω_{body} is $(R + H + B)\,\omega_{body}$. Hence the velocity of point B, the point of contact between the bat and ball, is

$$v_B = (R + H + B)\omega_{body} + (H + B)\omega_{wrists}$$

This means (combining the two immediately preceding equations) that

$$v_B = B(\omega_{body} + \omega_{wrists}) + v_{2b}$$

and this tells us that the motion of the point of contact B can be envisioned as the sum of two motions, one of which is the translational motion of the center of mass and the other a rotation of the bat about the center of mass with angular velocity $\omega_2 = \omega_{body} + \omega_{wrists}$ whereas before the subscript 2 stands for the bat.

Notice that if the point of impact were at the center of mass of the bat, v_B and v_{2b} would be the same (because B would be zero) regardless of the rotational velocity of the bat. If, on the other hand, the ball and bat do not collide at the center of mass of the bat, a change in the rotational velocity of the bat will occur. This change in angular velocity is the result of torque, just as the change in linear (translational) velocity in the equation

$$F = \frac{(v - v_0)m}{t}$$

resulted from a force. Suppose the force exerted on the bat during the collision is $-F_1$ and the force on the ball is F_1. The torque on the bat about its center of mass is $-BF_1$ and, equating this torque multiplied by the time of its application to the change in angular momentum of the bat, we find that

$$-BF_1 t = I_0(\omega_{2a} - \omega_{2b})$$

where ω_{2a} is the angular velocity of the bat (about its center of mass) after the collision and ω_{2b} is its angular velocity before the collision, ω_{body} + ω_{wrists} For the ball,

$$BF_1 t = Bm_1 (v_{1a} - v_{1b})$$

which is the same as equation

$$F_1 t = m_1 v_{1a} - m_1 v_{1b}$$

since we are here assuming that the collision occurs at the center of mass of the ball. Obviously most collisions will not occur at the center of mass of the ball, and, therefore, the ball will obtain rotational as well as translational kinetic energy. However, in typical bat-ball collisions the rotational kinetic energy is 5 to 10 percent of the translational kinetic energy. Therefore, neglecting the rotational kinetic energy of the ball is reasonable. Adding the equations

$$-BF_1 t = I_0 (\omega_{2a} - \omega_{2b})$$

and

$$BF_1 t = Bm_1 (v_{1a} - v_{1b})$$

and noting that the sum is zero because angular momentum is conserved during a collision, we find that

$$I_0 (\omega_{2a} - \omega_{2b}) + Bm_1 (v_{1a} - v_{1b}) = 0$$

The immediately preceding equation and the equation

$$m_1 v_{1b} + m_2 v_{2b} = m_1 v_{1a} + m_2 v_{2a}$$

are both true during a bat-ball collision, as is the equation

$$e = \frac{\text{relative speed after collision}}{\text{relative speed before collision}}$$

However, the equation

$$e = -\frac{v_{1a} - v_{2a}}{v_{1b} - v_{2b}}$$

is not the correct expression for this model because the relative velocities there represent those of the centers of mass of the ball and bat and not the velocities at the point B. where the collision takes place. The correct expression for this model is

$$e = -\frac{v_{1a} - v_{2a} - B\omega_{2a}}{v_{1b} - v_{2b} - B\omega_{2b}}$$

The equations

$$m_1 v_{1b} + m_2 v_{2b} = m_1 v_{1a} + m_2 v_{2a}$$

$$I_0 (\omega_{2a} - \omega_{2b}) + BM_1 (v_{1a} - v_{1b}) = 0$$

$$e = \frac{v_{1a} - v_{2a} - B\omega_{2a}}{v_{1b} - v_{2b} - B\omega_{2b}}$$

can be solved simultaneously, and after a good deal of algebra we find that

$$v_{1a} = \frac{-v_{1b}\left[e - \dfrac{m_1}{m_2} - \dfrac{m_1 B^2}{I_0}\right] + (1 + e)(v_{2b} + B\omega_{2b})}{1 + \dfrac{m_1}{m_2} + \dfrac{m_1 B^2}{I_0}}$$

Comparing the above equation to the following translation only equation

$$v_{1a} = \frac{(m_1 - em_2) v_{1b} + (m_2 + em_2) v_{2b}}{m_1 + m_2}$$

shows that the two are identical when $B = 0$, as they should be. This equation has been previously derived by Brancazio[3,15] (using different symbols). He also gives a lengthy and interesting discussion of what the equation can tell us about the effects on hitting of the various values in the equation. Of particular interest is the variation of v_{1a} the batted-ball speed, with B, the distance from the center of mass of the bat to the collision point, and with $\sqrt{\dfrac{I_0}{m_2}}$ a measure of the mass distribution, or shape, of the bat. Brancazio has calculated the values of $\sqrt{\dfrac{I_0}{m_2}}$ which is called the *radius of gyration* about the center of mass for various bats. Some of his data are presented in Table 7. The W1 refers to a typical wood bat; A1, to an aluminum bat with a wall thickness of 0.1255 inch that has the same shape and weight as W1; an A2 to an aluminum bat of the same shape as A1 but with a wall thickness of 0.1 inch. From this table, we see that a 31-oz, 34-in wooden bat has a smaller radius of gyration than either a 31-oz or a 25-oz aluminum bat of equal length. Brancazio says, "In general . . . it is to the batter's advantage to select a bat with a large value of k_0." Next, Brancazio calculated the value of B that resulted in the maximum value of v_{1a}. This is the collision point for which the bat imparts maximum energy to the ball. He calls it the *maximum energy transfer* or *MET point.* It is important to note that the MET point is not, as is commonly believed, the center of percussion of the bat.

Let us first investigate how v_{1a} varies with the collision point for two bats of equal weight, one a wooden bat and the other an aluminum bat. In each case, we assume the total bat kinetic energy before the collision is 300 foot-pounds per second (or 408 joules), which, as is shown in Table 5, is the largest value recorded for the San Francisco Giant players. The results are shown in Figure 47. There are several interesting features in these two curves. First, the speed of the ball leaving the bat is about 120 mph, which is greater than home run speed. Second, the MET point is closer to the handle for the aluminum bat than for the wood bat. Finally, and perhaps the most important point of all, the curve is significantly broader for the aluminum bat. As pointed out by Brancazio, this has some important implications regarding hitting with wood and aluminum bats.

The MET point for an aluminum bat is closer to the handle than the MET point for a similar wood bat. More important, the effective hitting surface is slightly shorter toward the barrel end, and much longer near the handle for an aluminum bat. This means that alu-

Table 7 Properties of Some Bats			
	W1	**A1**	**A2**
Length (in)	34	34	34
Weight w (oz)	31	31	25
Distance from the end of the handle to the center mass (in)	21.9	19.6	19.5
Moment of inertia about the center of mass, I_0 (lb-ft^2)	0.0354	0.0432	0.0353
Radius of gyration k_0 (in)	9.2	10.2	10.2

minum bats are much more effective for hitting inside pitches. Those who have hit with both aluminum and wood bats report that this is true. The principal reasons for this are that k_0, the radius of gyration about the center of mass, is larger for aluminum bats and the center of mass itself is closer to the handle. Brancazio has performed a similar analysis for corked bats and reported that corking the bat does not help. It would appear, however, that there might be some bat design applicable to wooden bats that would increase the value of k_0 while still maintaining the general requirement that the bat be cylindrical and made of solid wood. The best shape of a wooden bat has almost certainly not yet been determined.

Will Outlawing the Aluminum Bat Will Endanger Pitchers?

Recently some data published on the Internet compared the performance of aluminum and wood bats of the same weight and length. The author suggested outlawing aluminum bats because the batted-ball speed (the speed of the ball after the bat-ball collision) might be higher for the aluminum bat. However, this comparison may be a red herring concerning whether outlawing the aluminum bat would make it safer or more dangerous for pitchers. These data are irrelevant, because if the aluminum bat was outlawed almost no one would switch from a 32-oz aluminum

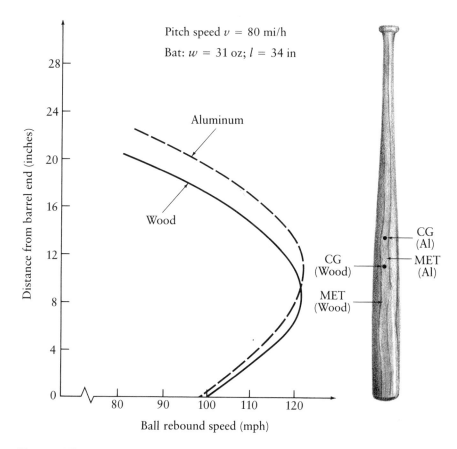

Figure 47   Batted-ball speed as a function of the point of impact for a wooden bat and an aluminum bat. The effective hitting surface is larger and closer to the handle for the aluminum bat. (Adapted from P. Brancazio, 1987.)

bat to a 32-oz wood bat. Almost all switches would be from lighter aluminum bats to heavier wood bats. This would produce higher batted-ball speed in almost every case.

Figure 43 shows the bat speed and the batted-ball speed for a typical major league baseball player. The given bat speeds are the speed of the center of mass of the bat. Our calculations assume a fastball with a speed of 90 mph and 1000 rpm of topspin, a contact point 3 feet above the ground, a contact angle of 30 degrees up, a batted-ball spin of −1000 rpm and a coefficient of restitution of 0.55.

For the data of Figure 43, a 27-oz bat would produce a batted-ball speed of 113 mph, a 29-oz bat would produce a batted-ball speed of 114 mph, and a 32-oz bat would produce a batted-ball speed of 115 mph.

In the bat weight range from 27 to 32 ounces, the batted-ball speed curve slopes upward, which means that as the bat weight increases the batted-ball speed increases. This statement is true for all 28 of the San Francisco Giants and for 19 of the 28 University of Arizona softball players that we measured. Therefore, for most batters, outlawing aluminum bats would produce faster batted-ball speeds.

Oblique Collisions: Putting Spin on the Ball

Until now, we have dealt only with a head-on collision, wherein the bat and ball are moving in precisely opposite directions before and after the collision. Motion occurs in only one direction. We now wish to generalize our discussion to include the possibility that the bat strikes the ball obliquely, as shown in Figure 48. We again treat the bat as a cylinder whose axis is perpendicular to the path of the ball, but this time the collision might take place as a glancing blow. In addition, we allow for the possibility that the baseball might be rotating before it strikes the bat, and will almost certainly be rotating afterward. The theory of oblique collisions is complicated, and we will not go into its details, which may be found in Werner Goldsmith's book *Impact*.[17] Even in a simple treatment of oblique collisions, however, we must deal with the possibility that the surface of the ball might slip relative to the wooden surface of the bat. We therefore introduce a new property, the *coefficient of friction*.

Imagine a block resting on a flat surface. If both the surface and the block are perfectly smooth, as for an ice cube on a glass tabletop, the tiniest sideways force will cause the block to slide across the surface (recall Figure 14). For real surfaces, there is always a frictional force that must be overcome before the block will slide (recall Figure 15). This force is equal to the coefficient of friction, u, times the weight of the block:

$$\text{Force} = \mu \times \text{weight}$$

If the block is pushed sideways with a force smaller than this value, it will not move. If it is pushed with a larger force, it will slide along the surface.

When a baseball collides with a bat obliquely, there is a force on the ball whose direction goes along a line connecting the centers of the bat

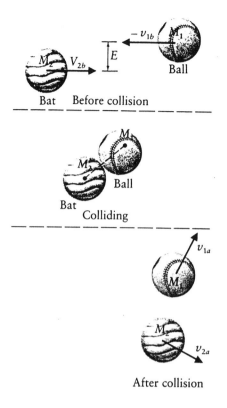

Figure 48   A ball and bat involved in an oblique collision.

and the ball. This is the force that, acting through the coefficient of restitution, causes the ball to bounce off the bat. It also acts like the weight force in the equation for the coefficient of friction. There is, in addition, a force on the ball perpendicular to the line connecting the bat and ball centers. It results from the frictional force between the bat and the ball as they tend to slide past one another. If the ball strikes the bat head on, the frictional force will be zero. As the collision becomes increasingly oblique, the force along the ball and bat centers becomes smaller and the frictional force becomes larger, imparting a larger and larger spin to the ball as it leaves the bat. Finally, however, the frictional force reaches some maximum value (depending on the coefficient of friction and how hard the ball is hit). When the collision becomes more oblique than this, the ball surface slips relative to the bat surface and the frictional force reaches its maximum value: that given by the immediately preceding equation.

We have performed calculations when both the bat speed and the pitch speed were 72 mph. This would be a slow pitch by a major leaguer (see Table 1) and a fast swing (see Table 2). Calculations have been made using three values of the coefficient of friction: 0.1, 0.075 and 0.05. Results are shown graphically in Figures 49 and 50, and in Tables 8, 9 and 10 for precollision ball rotation rates of 1900 rpm, 0 rpm, and −1900 rpm, approximately corresponding to the outside ranges for an overhand curve ball (drop), a knuckleball, and a fastball, as described in Table 1.

To simplify matters somewhat, let us assume the batter swings in such a way that the direction of the bat velocity is precisely opposite the direction of the ball velocity. The geometry of the situations just a moment before the collision and a few moments after the collision are shown in Figure 48. We will call E the *error in the batter's swing* (as shown in Figure 48).

Consider first the curve ball where $\omega = +1900$ rpm. For small values of E, the distance by which the batter undercuts the ball, the ball does not slip relative to the bat and the rotation rate of the ball after the collision, called the *batted-ball rotation rate*, increases with E. If the coefficient of friction is 0.1, the ball surface does not slip relative to the bat until E reaches 0.7 inch, at which point the ball leaves the bat with a backspin rate of 7900 rpm at a launch angle 15 degrees above horizontal. The batted-ball speed is 120 mph, only slightly smaller than the value when $E = 0$, which is 124 mph. Undercutting the ball by a greater amount leads to slip between the ball and bat surfaces. However, even then, both batted-ball speed and rotation rate decrease only very slowly. At $E = 1.0$ inch, the batted-ball rotation rate is 7677 rpm, the speed is 116 mph and the launch angle is 26.5 degrees.

For the fastball ($\omega = -1900$ rpm), the maximum batted-ball rotation rate when $\mu = 0.1$ is 4190 rpm at $E = 0.46$ inch. The batted-ball speed is 123 mph at a launch angle of 6.6 degrees. If $E = 1$ inch, the speed has decreased to 116 mph, the spin rate is 3858 rpm and the launch angle is 26.5 degrees.

When the coefficient of friction is lower, the critical value of E is reached sooner. For cases when slip occurs, the rotation rates are lower and the launch angles are larger. Postcollision ball speeds are hardly affected.

There are important lessons to be learned here. When the batter undercuts the ball, even by as much as an inch, very little of the batted-ball speed is lost. The ball can attain a large backspin, even if the ball and bat surfaces slip relative to one another. The launch angle increases at a

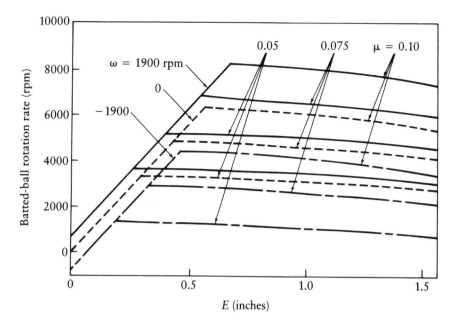

Figure 49   The batted-ball rotation rate (backspin) as a function of E, the distance between the velocity vector of the center of the bat and the velocity vector of the center of the ball as shown in Figure 48.

moderate rate until the ball and bat surfaces slip relative to one another, after which it increases very rapidly. High launch angles, of course, mean pop-ups or foul-backs.

Apparently, the batter has more margin for error than conventional wisdom implies. Undercutting the ball by one half to three quarters of an inch gives the ball backspin while hardly sacrificing batted-ball speed and without popping the ball up. Keep in mind that Steve Garvey said the ball carries better if you undercut it a little (as mentioned in the last paragraph of Chapter 4). Evidently, he knows what we have just proved. In the next chapter we study the motions of batted balls after they leave the bat, and show how much farther the ball goes if it has backspin.

Notes

1. Ted Williams and John Underwood, *The Science of Hitting*, rev. ed., Simon and Schuster, New York, 1986. An instruction manual by one of the best hitters of all time.

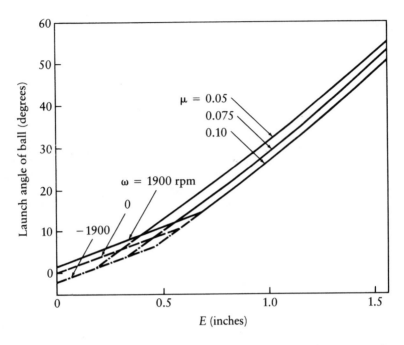

Figure 50   The launch angle of the ball as a function of the batter's error, E.

2. Paul Kirkpatrick, "Batting the Ball," *American Journal of Physics,* vol. 31, 1963, pp. 606–613. A theoretical physics analysis of bat-ball collisions.

3. Peter J. Brancazio, "Swinging for the Fences: The Physics of the Baseball Bat," paper presented at the New England section of the American Physical Society meeting, October 1987. The physics of bat-ball collisions.

4. Tom Meany, "Why So Many Home Runs—Is It $E = \frac{1}{2} mV^2$?," *Collier's,* September 28, 1956.

5. J. Kaat, "Foul Ball," *Popular Mechanics,* vols. 142 and 145, May 1988, pp. 82–85. Illustrations of some naughty things baseball players have done to their bats.

6. D. Gutman, "The Physics of Foul Play," *Discover,* April 1988, pp. 70–79. A light treatment of some things baseball players have done to their bats.

7. W. O. Fenn and B. S. Marsh, "Muscular Force at Different Speeds of Shortening," *Journal of Physiology,* vol. 85, 1935, pp. 277–297. A classic paper on the muscle force-velocity relationship.

8. A. V. Hill, "The Heat of Shortening and Dynamic Constraints of Muscle," *Proceedings of the Royal Society, London (ser. B),* vol. 126, 1938, pp. 136–195.

Table 8   Collision Results for 1900-rpm Precollision Rotation Rate

μ	E (in)	v (mph)	Launch Angle θ (deg)	ω (rpm)
0.1	0	124	1.9	700
	0.2	124	5.4	2800
	0.4	123	9.0	4800
	0.6	121	12.8	6900
	0.7*	120	15.0	7900
	0.8	119	18.2	7850
	1.2	114	33.6	7500
0.075	0	124	1.9	700
	0.2	124	5.4	2800
	0.4	123	9.0	4800
	0.55*	122	12.0	6470
	0.8	119	20.6	6370
0.05	0	124	1.9	700
	0.2	124	5.4	2800
	0.43*	123	10.2	4980
	0.8	120	23.0	4880

*Slip occurs at large E.

Original derivation of the Hill equation for the nonlinear force-velocity relationship of muscle.

9. D. R. Wilke, "The Relation Between Force and Velocity in Human Muscle," *Journal of Physiology,* vol. 110, 1950, pp. 249–280. A careful study of the force-velocity relationship of human muscle.

10. B. R. Jewell and D. R. Wilke, "The Mechanical Properties of Relaxing Muscle," *Journal of Physiology,* vol. 152, 1960, pp. 30–47. A study of the muscle force-velocity relationship of nonactivated muscle.

11. G. C. Agarwal and G. L. Gottlieb, "Mathematical Modeling and Simulation of the Postural Control Loop—Part 11," *CRC Critical Reviews in Biomedical Engineering,* vol. 11, 1984, pp. 113–154. A review of neuromuscular models.

12. A. T. Bahill and W. J. Karnavas, "Determining Ideal Baseball Bat Weights Using Muscle Force-Velocity Relationships," *Biological Cybernetics,* vol. 62, 1989, pp. 89–97. The original ideal bat weight study.

Table 9   Collision Results for −1900-rpm Precollision Rotation Rate

μ	E (in)	v (mph)	Launch Angle θ (deg)	ω (rpm)
0.1	0	124	−1.9	−700
	0.2	124	1.7	1380
	0.4	123	5.2	3460
	0.46*	123	6.6	4190
	0.8	119	18.2	4040
0.075	0	124	−1.9	−700
	0.2	124	1.7	1380
	0.32*	124	3.8	2630
	0.4	123	6.4	2700
	0.8	119	20.6	2550
	1.2	115	35.9	2300
0.05	0	124	−1.9	−700
	0.18*	124	1.4	1180
	0.4	123	8.85	1160
	0.8	119	23.0	1060

*Slip occurs at large E.

13. A. T. Bahill and W. J. Karnavas, "The Ideal Baseball Bat," *New Scientist*, 130, (no. 1763), 6 April 1991, pp. 26–31. The ideal bat weight study translated into British.

14. A. T. Bahill and M. Morna Freitas, "Two Methods for Recommending Bat Weights," *Annals of Biomedical Engineering*, 23, 1995, pp. 436–444, 1995. A database analysis to find the best bat weight without using a specialized instrument such as the Bat Chooser.

15. Peter J. Brancazio, "Optimal Bat Weight," unpublished manuscript. The physics of bat-ball collisions with suggestions of how the bat could be changed to benefit the batter.

16. H. Brody, "The Sweet Spot of a Baseball Bat," *American Journal of Physics*, vol. 54, 1986, pp. 640–643.

17. Werner Goldsmith, *Impact: The Theory and Physical Behaviour of Colliding Solids*. E. Arnold Publishing, London, 1960. How things behave when they collide.

Table 10	Collision Results for 0-rpm Precollision Rotation Rate			
μ	E (in)	v (mph)	Launch Angle (deg)	ω (rpm)
0.1	0	124	0	0
	0.2	124	3.5	2090
	0.4	123	7.1	4160
	0.6*	121	11.0	6060
	0.8	119	18.2	5950
	1.2	114	33.6	5620
0.075	0	124	0	0
	0.2	124	3.5	2090
	0.4	123	7.1	4160
	0.43*	123	7.9	4560
	0.8	119	20.6	4460
	1.2	115	35.9	4210
0.05	0	124	0	0
	0.2	124	3.5	2090
	0.29*	123	5.3	3080
	0.4	123	8.8	3070
	0.8	120	23.0	2970

*Slip occurs at large E.

Also of interest: Robert K. Adair, The Physics of Baseball, 2d ed. New York, Harper Perennial, & Row, 1994. This book summarizes many experiments that Adair performed as "physicist for the National League."

L. L. Van Zandt, "The Dynamical Theory of the Baseball Bat." *American Journal of Physics*, 60, no. 2 1992, pp. 172–181. A theoretical study of the baseball bat.

The Society for Baseball Research (SABR) is a not-for-profit corporation of baseball researchers and enthusiasts who collect, organize and disseminate baseball information. They publish a journal and many books. SABR, 812 Huron Road East, Suite 719, Cleveland, OH 44115, (216) 575-0500.

The Flight of the Ball:

The Kinetics of Fly Balls

K inetics is the study of the relation between the forces acting on a body, the mass of the body and the resulting motion of the body. We have already done some kinetics calculations in Chapters 2 and 4. We will now study in somewhat detail the motion of a baseball in response to the forces it normally experiences in flight.

The equation that is central to our study is the quantitative expression of Newton's second law of motion:

$$F = ma$$

There are a number of forces acting on the ball. Obviously, the force exerted by gravity on the ball (the weight) will pull the ball toward the ground. If the ball is spinning, a "lift" force will be exerted in a direction mutually perpendicular to the path of the ball and the axis of spin, as explained in Chapter 2. In that same chapter, we mentioned another force: the "drag" force due to the air rushing past the surface of the ball. As a first approximation, we neglected the effect of the drag force on the path of a pitched ball. This is a good approximation as long as the ball is not in the air for longer than about half a second, as happens for pitched balls. (Even then, certain pitches, notably the knuckleball, are thought to be affected in an important way by the drag force. We will come back to this later.) Fly balls, which might be airborne for 5 to 10 seconds, are strongly affected by both aerodynamic drag and lift induced by the spin of the ball.

"HERE – IF YOU'RE SO DAMN SMART, YOU PITCH A FEW INNINGS."

As long as the force on the ball is only that caused by gravity, the ball's path will be described by the parabola, as we discussed in Chapter 2, where we first introduced kinetics. In fact, in that chapter we computed the range R of a ball, or the distance a ball will travel when thrown or batted with given initial velocity. The range is given by the equation

$$R = \frac{2v_{x0}\,v_{z0}}{g}$$

We stated in Chapter 5 that if a ball is to travel far enough to be a home run, it needs to leave the bat at about 110 mph (161 ft/s), and in Table 4 we showed that many members of the San Francisco Giants could produce such batted-ball speeds. The range, as computed from the immediately preceding equation, will be a maximum if the ball is hit at a 45-degree angle from the horizontal. In this case, the initial velocities v_{x0} and v_{z0} are both 114 feet per second. (Note that $\sqrt{v_{x0}^2 + v_{z0}^2} = $ 161 ft/s.) The maximum range is

$$R_{max} = \frac{2\,(114)\,(114)}{32} = 812 \text{ ft}$$

It is improbable that anyone has hit a baseball this far.

The trouble with the above calculation is that we assumed that gravity was the only force acting on the ball. In truth, both aerodynamic drag and lift exert quite large forces. We have already discussed the lift force caused by the spin of the ball. We need to discuss the drag force before proceeding to calculate the trajectories of fly balls.

Drag Force: The Drag Coefficient

As in this case of the lift force, the drag force F_D on a spinning ball is best related to other variables through the set of three dimensionless group equations introduced in Chapter 4. These are

$$G_1 = C_D = \frac{2F_D}{\rho A v^2}$$

$$G_2 = SP = \frac{R\omega}{v}$$

$$G_3 = \mathrm{Re} = \frac{2vR}{\nu}$$

This time, because of the higher velocity, the *drag coefficient* C_D depends on both the spin parameter *SP*, and G_3, which is called the *Reynolds number* and is written *Re*. Unfortunately, the variation of C_D with *Re* is rather complex. Let us examine what is currently known about this variation.

The maximum value of the Reynolds number that we might expect in baseball would correspond to a ball speed of about 110 to 115 mph, say 170 ft/s. The kinematic viscosity v of air at a temperature of 80°F is about 16×10^{-5} ft²/s. Under these conditions, a 2.85-in diameter ball would have a Reynolds number of

$$Re = \frac{2\,(170)\,(2.85/12)}{16 \times 10^{-5}} = 2.5 \times 10^5$$

Typical Reynolds numbers experienced by baseballs in play are large, with a maximum of around 2.5×10^5. This is just about the range of Reynolds number where a peculiar phenomenon takes place, leading to a very strong and nonlinear variation of *CD* with *Re*.

Figure 51 shows the variation of drag coefficient with Reynolds number for nonspinning spheres as measured by E. Achenbach.[1] There are several curves, because there is actually another dimensionless group, called the *relative roughness*, that enters the picture. The relative roughness is defined as ε/D, where D is the diameter of the sphere and ε is the height of the fine grains of sand that were evenly distributed over the surface of the balls to make them rough in experiments leading to Figure 51. The major facts to notice about Figure 51 are that (1) the drag coefficient is nearly constant over a range of Reynolds numbers from about 10^4 to just over 10^5, (2) after some critical Reynolds number is exceeded, the drag coefficient drops sharply by a factor of 4 or more, and then begins to increase slowly and (3) the value of the critical Reynolds number depends on the relative roughness.

The sharp drop in the drag coefficient is not difficult to explain. Refer back for a moment to Figure 18. When a fluid flows past a sphere, the boundary layer tends to separate and a wake forms behind the ball. The pressure within the wake region is much lower than the pressure on the

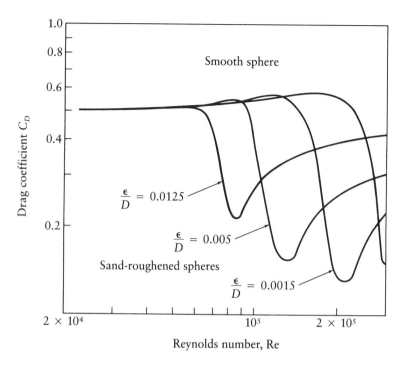

Figure 51   *The drag force (as represented here by the dimensionless drag coefficient) is not constant, but varies with the speed of the ball (as represented here by the dimensionless Reynolds number). For very smooth spheres, it is nearly constant up to speeds of about 40 mph, above which it drops off quickly. For sand-roughened spheres, it begins to decrease at much smaller ball speeds, but does not decrease by quite so much. This is why golf balls are made with dimpled surfaces. Spheres with roughened surfaces will fly further than smooth spheres.*

forward half of the ball. At high Reynolds numbers, this pressure difference accounts for most of the drag force on the ball. We further noted in Chapter 3 that C. Weisselsberger discovered that by placing a wire around the front part of the sphere one could introduce turbulence into the boundary layer and thereby delay separation and reduce the size of the wake, as shown in the lower picture in Figure 18. Since force is pressure times area, the reduced low-pressure wake area reduces the drag force.

What Figure 51 tells us is that the boundary layer will become turbulent if the Reynolds number is made large enough. The rougher the surface of the sphere, the lower the Reynolds number at which the drag coefficient drops.

The manufacturers of golf balls have known this for a long time. Many years ago, when golf balls were smooth, astute golfers noticed that old balls with cut and marred surfaces could be driven much further than the new, smooth balls. At first, balls with bumps distributed on the surface were introduced. These tended to get dirty and to retain lumps of mud. The present day dimpled-surface balls were then introduced.

Some writers (notably C. Frohlich[2]) have assumed that since the height of the stitches above the surface of a baseball is about 0.5 mm, the baseball behaves more or less like a rough sphere with a relative roughness of about 0.007. While few drag force measurements are available for baseballs, and none has been reported for spinning baseballs, the available data do not appear to indicate that a baseball behaves like the sand-roughened sphere used in Achenbach's experiments. Figure 52 shows some drag coefficient data reported by Richard Gonzalez[3] along with some golf ball data reported by P. W. Bearman and J. K. Harvey.[4] Gonzalez used nonspinning baseballs, and the limitations on the wind tunnel used allowed velocities of only about 75 mph. The huge drops in drag coefficient experienced by sand-roughened spheres do not appear to be present. However, J. W. Hollenberg,[5] also studying nonspinning balls, found large drops in drag coefficient at speeds between 50 and 80 mph for some orientations of the baseball.

The upshot of all this is that we simply do not know how the drag force on a spinning baseball varies with speed. Based on the data of Gonzalez and Hollenberg, we can guess that the drag coefficient decreases with increasing speed, but probably not nearly as rapidly as it does for the smooth or sand-roughened spheres studied by Achenbach. We would guess that the decrease is from about 0.6 to 0.4 over the range of Reynolds numbers typical of baseballs in play. (We note that things might be quite different for spinning balls.) We will make our trajectory calculations based on an assumed constant drag coefficient with a value of 0.5, cautioning the reader that this may be a crude approximation. *Comparative values* of the distance traveled by the ball under a variety of conditions should be reasonably represen-

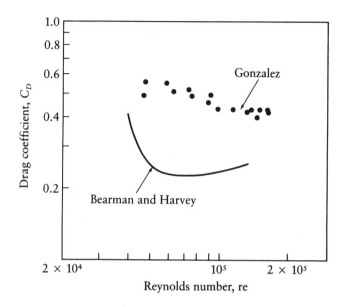

Figure 52   Drag coefficient as a function of Reynolds number for base-balls and golf balls. (Data collected by Richard Gonzalez at Tulane University and P. W. Bearman and J. K. Harvey of Imperial College.)

tative of a real-life situation. Frohlich has performed some calculations of the distance traveled by balls with drag coefficients varying as in sand-roughened spheres. He did not, however, account for the lift force caused by spin. In this chapter, we want to study this effect. The results that we will present have been published by Robert G. Watts and S. Baroni[6].

Trajectories

Determining the path of a ball with Newton's second law generally requires the use of a computer. We will describe the technique briefly and then present some results graphically in Figure 53. At some instant the ball is located at a point (x, z) and is moving with velocity (v_x, v_z). Let F_{DX} and F_{DZ} be the drag force components in the x and z directions. Similarly, F_{LX} and F_{LZ} are the lift force components. The quantity v is the speed, which is equal to $\sqrt{v_x^2 + v_z^2}$. Newton's second law yields two equations, one for the x-direction acceleration and one for the z-direction acceleration:

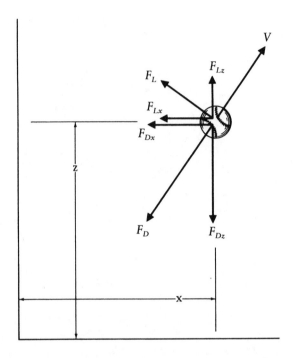

Figure 53 In addition to the force exerted on a ball by gravity, a real ball in flight is subject to at least two other forces, the aerodynamic drag force in a direction opposite the direction of motion and the lift force, mutually perpendicular to the direction of motion and the spin axis. In this figure (as well as the analysis in the text), we assume the spin axis of the ball is directed into the page. The x- and z-direction forces are obtained by resolving the lift and drag force into the upward and horizontal components.

$$ma_x = F_{Dx} - F_{Lx}$$

$$ma_z = -mg - F_{Dz} + F_{Lz}$$

If the speed v of the ball and its rotation rate are known, F_D and F_L can be calculated from

$$F_L = \frac{1}{2} \rho v^2 \pi R^2 C_L$$

and

$$F_D = \frac{1}{2} \rho v^2 \pi R^2 C_D$$

If v_x and v_z are known, the components of F_D and F_L can be determined from similar triangles:

$$\frac{v_x}{v} = \frac{F_{Dx}}{F_D} = \frac{F_{Lz}}{F_L}$$

and

$$\frac{v_z}{v} = \frac{F_{Dz}}{F_D} = \frac{F_{Lx}}{F_L}$$

We can solve the equations sequentially in the following way:

1. Suppose we know the values of v_x and v_z at time $t = 0$, v_{x0} and v_{z0}.
2. Using these velocities, we calculate all the terms on the right side of the equations.

$$ma_x = F_{Dx} - F_{Lx}$$

$$ma_z = -mg - F_{Dz} + F_{Lz}$$

3. Divide by the mass of the ball to obtain the acceleration terms a_x and a_z.
4. Acceleration is the rate of velocity change with respect to time: $a_x = (v_x - v_{x0}/\Delta t)$ and $a_z = (v_z - v_{z0}/\Delta t)$ where Δt is the change in time. In truth, this formula for acceleration gives the average acceleration over the time increment. If you make even a small error, there is sometimes enough difference between the average and instantaneous acceleration values to screw up the trajectory calculations. For you mathematics whizzes, we have used a third-order Runge-Kutta formulation in our numerical calculations. The current description gives the general idea though.

5. Now we know the new velocity components, so we can start from there and repeat things from Step 1. Meanwhile, the distance the ball traveled during the first Δt was

$$\Delta x = \frac{1}{2}(v_x + v_{x0})\Delta t$$

$$\Delta z = \frac{1}{2}(v_z + v_{z0})\Delta t$$

The trajectory equations were solved initially using time steps of 0.0003 and 0.03 s. An initial ball speed of 110 mph was used in all calculations.

Typically, the difference in ranges was about 2 feet, the smaller time increments giving the longer, more accurate ranges. Some typical trajectories are shown in Figure 54. The dashed line shows the path of a ball with no lift or drag forces at all. Other curves show paths with no lift, but realistic drag force, with topspin of 4000 rpm, and with backspins of 4000 rpm and 6000 rpm. In all cases, the launch angle is 30 degrees above the horizontal and the initial speed is 110 mph. Without drag, the

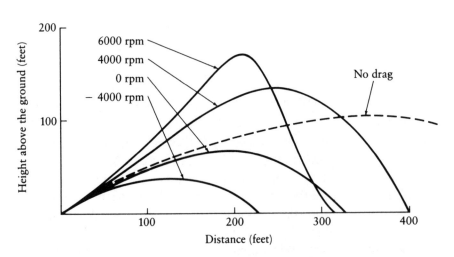

Figure 54 *Typical trajectories of batted balls. The ball speed leaving the bat is 110 mph, and the launch angle is 30 degrees above the horizontal.*

ball would travel 700 feet! Note that you can calculate this for yourself using the equation

$$R = \frac{2v_{x0}\,v_{z0}}{g}$$

The drag force has a large effect. When the drag force, but not the lift force, is included, the ball goes only about half as far. With 4000-rpm backspin, the resulting lift on the ball keeps it in the air longer, and it travels 400 feet! If it were possible to hit a ball with a launch angle of 30 degrees and with 6000 rpm backspin, it would eventually rise at such a steep angle that the lift force would prevent its forward motion and it would travel just over 300 feet. Conversely, topspin leads to a sinking line drive. Results that are more general are shown plotted in Figures 55, 56 and 57. We have used time increments of 0.03 seconds in the calculations.

First, consider Figure 55. When the spin rate on the ball is 0, there is no lift force. The maximum range is about 335 feet. These numbers are similar to those of Frohlich[2] and Peter J. Brancazio.[7] As the spin rate increases, the maximum range increases dramatically and occurs at smaller launch angles. When the spin rate is 4000 rpm, the maximum range is 430 ft and the optimum launch angle is 18 degrees.

The data are displayed in a different manner in Figure 56. For a launch angle of 40 degrees, the spin rate giving the maximum range (361 ft) is 2300 rpm. If the launch angle is 20 degrees, a maximum range of 440 ft is reached with an optimum spin rate of 4780 rpm.

To plan an optimum strategy, the analytically inclined batter should study Figure 57. Curves of constant range are plotted against rotation rate and launch angle. Clearly, the best strategy is to obtain a high backspin rate on the ball at a low launch angle. The *combination* is important. Suppose, for example, that a batter launched a ball optimally just under 40 degrees with a spin rate of 0. The ball would travel 335 ft. If he launched the ball at 39 degrees with a spin rate of 4100 rpm, the range would again be 335 ft. However, if he attained a spin rate of 4100 rpm and launched the ball at the optimum angle of 16 degrees, the range would be 436 ft., 101 ft farther than the nonspinning ball.

These are spectacular numbers. Baseball is supposed to be a game of inches, not tens to hundreds of feet. A central question must be: How much backspin can one hope to attain while hitting a baseball without fouling the ball back? This question, of course, forces us to return our

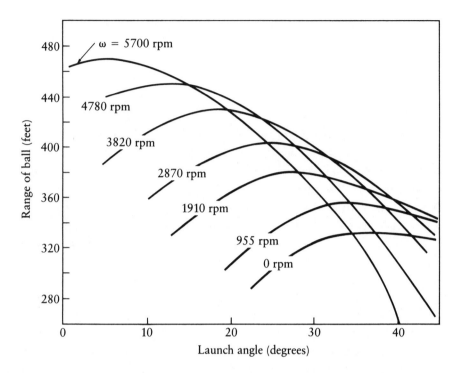

Figure 55 *This figure shows how the range of a batted baseball is affected by the spin rate and the launch angle. The speed of the ball leaving the bat is assumed to be 110 mph. To find the range of a particular ball, locate the launch angle on the horizontal axis. Trace upward along a vertical line until you reach a curve labeled with the spin rate of the bat Then trace along a horizontal line until you reach the vertical axis on the left and read the range.*

attention to the collision calculation. A typical rotation rate of a very good curve ball is about 2000 rpm. It is a little difficult to believe that a batter could produce a rotation rate of 8000 rpm on a fly ball, as is implied by our calculations for the case of a batter undercutting a curve ball by between 0.67 and 1 inch when the coefficient of friction is 0.1. It is also difficult to believe that the coefficient of friction between a cowhide sphere and a wooden cylinder is much smaller than 0.1. The coefficient of friction between two billiard balls has been reported as 0.04 by W. Muller.[8] It is possible, of course, that very large rotation rates can be imparted to baseballs. As far as we know, there

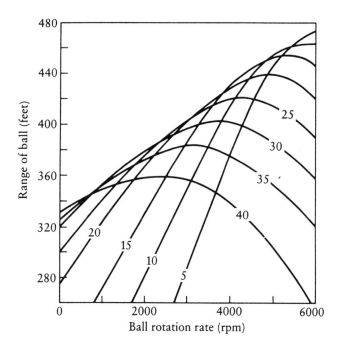

Figure 56 *This set of graphs gives the same information as Figure 54. Read in the rotation rate of the ball along the horizontal axis and trace vertically upward to the curve giving the appropriate launch angle. From this point, trace a horizontal line to the vertical axis and read the range.*

have been no rotation rate measurements of batted balls reported in the literature.

There are many potential problems in our collision analysis that are perhaps not made clear enough in the textbooks to students of physics. The classical theory of collision, which we used in our calculations, is called *stereomechanics* (Goldsmith[9]). It is based primarily on the laws of momentum for rigid bodies. The theory fails to account for local deformations at the contact point, and further assumes constant values for the coefficients of restitution and friction. The law of conservation of momentum must, of course, not be violated. It is probably the assumptions involving constant coefficients of restitution and friction that lead to problems. On the other hand, perhaps our striking predictions are good approximations of what happens in the real world. Obviously, a controlled experiment is in order. We wish we had the facilities to do such an experiment.

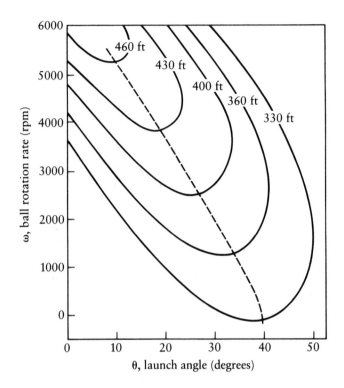

Figure 57 *From this figure, we can determine the best launch angle for a given ball backspin rate. The lines of constant range are like the contour lines on the map of a hill. The height of the hill (or the range of the ball) increases as you go toward the upper left-hand corner of the figure. The dashed line is like a ridge on the hill. Suppose a ball is launched with zero backspin. Start at ω = 0 and trace across to the dashed line. The ball should be launched 40 degrees above the horizontal and its range will be a little more than 330 ft. If the batter can manage to put 3000-rpm backspin on the ball, it should be launched at an angle of about 23 degrees above the horizontal, and its range would be about 410 ft.*

Some Tentative Conclusions About Hitting

The *qualitative* conclusion that one reaches from the trajectory calculations is unequivocal. To achieve maximum range, the backspin rate should be high and the launch angle should decrease as the spin rate increases. The effect of increasing the backspin rate is large. A modest backspin rate of 2000 rpm produces an increase in range of about 50 ft (when hit at an optimum launch angle) over that of a nonspinning ball.

Let us assume that the collision calculations in Chapter 5 are accurate in form and orders of magnitude. They imply that a batter should undercut the ball by 0.5 to 1 in. When the ball is undercut by more than this, little of the ball velocity or spin rate is lost, at least until the distance the ball is undercut (E) is greater than about 1.5 inches. The launch angle increases rapidly, however, and the advantage of having a large lift force due to large backspin is lost after about $E = 1$ in. Many writers in the past have suggested that hitting a baseball successfully is so difficult because the margin for error is so small. Conventional wisdom has it that an error in E of only 1 to 2 mm can mean the difference between a home run and a pop fly. Our calculations imply that this is not true. We show this with a possible scenario.

When an overhand curve ball crosses home plate, it is moving slightly downward at an angle of perhaps 10 degrees from the horizontal. Assume that the batter swings upward at the same angle. If he hits the ball flush ($E = 0$), and he has a bat speed of 60 mph, then from Figure 43, he will produce a batted-ball speed of 110 mph and the result will be a line drive. However, if he undercuts the ball by a quarter inch, the launch angle would be 5 degrees + 10 degrees = 15 degrees; the batted-ball speed would still be 110 mph, but this time the spin rate would be 2800 rpm. According to Figure 56 this ball would travel about 380 ft. An undercut of a half inch would send the ball about 440 ft, provided the coefficient of friction $\mu \geq 0.05$, while undercutting by three quarter inch would result in a 390-ft fly, if, $\mu > 0.075$. When $E = \frac{1}{2}$ in, approximately optimum conditions are produced, but changing E by a millimeter still produces a long hit. It might be argued, however, that a difference of $\frac{1}{4}$ to $\frac{3}{4}$ inch reduces the distance by a substantial 50 ft. The point we wish to make is that the ball has been hit solidly in each of the cases.

The final point we wish to make concerns the importance of the friction coefficient. Figure 49 implies that it is difficult to put large backspin on a fastball unless the coefficient of friction is reasonably large. Note also that a fastball should be undercut by a larger amount than a curve ball, provided the friction coefficient can be made reasonably high. There are many ways to accomplish this. Hitters have been known to carve grooves in their bats, and George Brett of the Kansas City Royals once had a home run called back because his bat had pine tar on it beyond the "legal" limit of 18 inches of handle. The American League president reinstated the home run later and stated that the rule was merely meant to keep the ball clean. We suspect George Brett knew perfectly well what he was doing.

Home Run Hitting and the Weather

The year 1987 seems to have been a banner year for home-run hitting. As we mentioned in Chapter 5, it is not likely that this increase was caused by an increase in the coefficient of restitution of the ball.

Recently several players have claimed that the spring and summer of 1987 were unusually hot and humid and that under these weather conditions the ball goes further. Increasing either the air temperature or humidity would have the effect of lowering the density of the air. This would in turn lower both the drag force and the lift force. While we do not intend to examine this question fully, we will present a few examples with the intention of determining the order of magnitude of the possible effect of weather on the range of batted balls.

We assume the air pressure is 14.7 pounds per square inch, which is normal atmospheric pressure at sea level. If the temperature is 50° F and the humidity is 0, the air density, computed from the ideal gas law, is 0.0779 pound per cubic foot. On the other hand, the density of 90° F air at 100 percent relative humidity is 0.0709 pound per cubic foot, a decrease of about 9 percent. This is an extreme case of density variation. We decreased the density term in the equations

$$F_L = \frac{1}{2} \rho v^2 \pi R^2 C_L$$

$$F_D = \frac{1}{2} \rho v^2 \pi R^2 C_D$$

by 10 percent and recomputed a few of the trajectories. The results are shown in Table 11. The launch angle is θ; the other terms are obvious. The first two rows keep the spin rate and launch angle constant, and vary the air density between its extremes. The last two rows keep the spin rate and launch angle constant at different values, while once again varying the air density between its extremes. Both pairs show that a 10 percent decrease in air density produces a 5 percent increase in how far the ball goes.

Table 11

Spin rate (rpm)	Launch angle (deg)	Temperature (°F)	Humidity (%)	Range (ft)
2865	35	50	0	383
2865	35	90	100	403
4775	20	50	0	439
4775	20	90	100	465

The decrease in air density decreases both the drag and the lift forces. Evidently, the decreased drag has the more important effect. A 10 percent decrease in air density causes the ball to travel 20 to 25 feet farther. This is, of course, a very large decrease in air density, and reflects extreme differences in atmospheric conditions. Nevertheless, the results show that temperature and humidity can potentially play an important role in home-run hitting.

In summary, on a hot, humid day air density is smaller than on a cold, dry day. This allows the ball to travel 5 percent farther. Temperature and humidity also affect the ball. Balls stored in 100 percent humidity absorb water, become heavier, have lower coefficients of restitution and subsequently fall somewhat short of balls stored in a normal environment. Conversely, balls stored in a freezer also have lower coefficients of restitution[10] for as long as the center is frozen and they fall short of balls at ambient temperature. Unless we know all the details, we should be careful in making generalizations about how the weather affects how far the ball goes.

Notes

1. E. Achenbach, "The Effect of Surface Roughness and Tunnel Blockage in the Flow Past Spheres," *Journal of Fluid Mechanics*, vol. 65, 1977. Drag forces on spheres.

2. C. Frohlich, "Aerodynamic Drag Crisis and Its Possible Effect on the Flight of

Baseballs," *American Journal of Physics,* vol. 52, April 1984. How the roughness of a baseball might affect how far it can be hit.

3. Richard Gonzalez, *Aerodynamics of a Knuckleball,* honors thesis in mechanical engineering, Tulane University, 1968. Wind tunnel measurements of the forces on the knuckleball.

4. P. W. Bearman and J. K. Harvey, "Golf Ball Aerodynamics," *Aeronauts Quarterly,* vol. 27, 1976, p. 112. Drag and lift forces on spinning golf balls by some British scientists.

5. J. W. Hollenberg, "Secrets of the Knuckleball," *The Bent of Tau Beta Pi,* series 4, vol. 77, 1986, pp. 26–30. More measurements, and computer predictions of the path of the knuckleball.

6. Robert G. Watts and S. Baroni, "Baseball-Bat Collisions and the Resulting Trajectories of Spinning Balls, *American Journal of Physics,* vol. 57, January 1989. How spin affects how far a baseball can be hit.

7. Peter J. Brancazio, *SportScience: Physical Laws and Optimum Performance,* Simon and Schuster, New York, 1984. A great book on the physics of all sports. A must for scientific sports fans.

8. W. Muller, *Ost. Ing. Arch.,* vol. 6, 1952. More about collisions.

9. Werner Goldsmith, *Impact: The Theory and Physical Behaviour of Colliding Solids,* E. Arnold Publishing, London, 1960. How things behave when they collide.

10. Robert K. Adair, *The Physics of Baseball,* 2d ed. New York, Harper Perennial, 1994. This book summarizes many experiments that Adair performed as "physicist for the National League."

Baseball Players Cannot Keep Their Eyes on the Ball[1]

Ted Williams, arguably the best hitter in the history of baseball, has described hitting a baseball as the most difficult single act in all of sports.[2] The velocity of a pitch approaches 100 mph. Tracking such a ball as it crosses the plate would require head and eye rotations in excess of 1000 degrees per second: a century of scientific literature says that humans cannot track targets moving faster than 90 degrees per second. Yet, major league batters manage to hit the ball with force consistently and are able to "get a piece of the ball" in 80 percent of their swings.[3] In this chapter we investigate how they do this by examining the head and eye movements of a professional athlete tracking a pitched ball, and we demonstrate the superiority of his eye movements and head-eye coordination to those of college athletes.

We studied our batter's eye movements. The purpose of the eye movement system is to keep the fovea, the region of the retina with the greatest visual acuity, on the object of interest. To accomplish this task, the following four types of eye movements work in harmony: *saccadic eye movements*, which are used in reading text or scanning a roomful of people; *vestibulo-ocular eye movements*, used to maintain fixation

(171)

during head movements; *vergence eye movements,* used when looking between near and far objects; and *smooth-pursuit eye movements,* used when tracking a moving object. These four types of eye movements have four independent control systems, involving different areas of the brain. Their dynamic properties, such as latency, speed and bandwidth are different, and they are affected differently by alcohol, fatigue, drugs and disease.

The specific actions of these four systems can be illustrated by the example of a duck hunter sitting in a rowboat on a lake. He scans the sky using saccadic eye movements, jerking his eyes quickly from one fixation point to the next. When he sees a duck, he tracks it using smooth-pursuit eye movements. If the duck lands in the water right next to his boat, he moves his eyes toward each other with vergence eye movements. Throughout all this, he uses vestibulo-ocular eye movements to compensate for the movement of his head caused by the rocking of the boat. Thus, all four systems are continually used to move the eyes.

In addition to each of these eye-movement systems, the batter has the potential to use the head-movement system. Earlier studies[1] suggested three increasingly more complex strategies for tracking the baseball: track the ball with only eye movements and fall behind in the last 10 ft; track the ball with head movements and smooth pursuit eye movements and fall behind in the last 5 ft; track the ball over the first part of its trajectory with smooth pursuit eye movements, make a saccadic eye movement to a predicted point ahead of the ball, continue to follow it with peripheral vision and, finally, at the end of the ball's flight, resume smooth pursuit tracking with the ball's image on the fovea. We will examine each of these strategies.

The Simulated Fastball

To discover how well a batter tracks the ball, we had to be able to determine the position of the ball at all times, and thus we could not use a real pitcher or a throwing machine. Instead, we simulated the trajectory of a pitched baseball (see Figure 58). We threaded a fishing line through a white plastic ball and stretched this line between two supports, which were set 80 ft apart in order to accommodate the 60.5 ft between pitcher and batter; a string was attached to the ball and wrapped around a pulley attached to a motor, so that when the motor was turned on, the string pulled the ball down the line at speeds between 60 mph and 90 mph. The ball crossed the plate 2.5 ft away from the subject's shoulders, simulating a high-and-outside fastball thrown by a left-handed pitcher to a right-handed batter.

Figure 58   *Our simulated fastball was produced when the white plastic ball was pulled along the fishing line by the string, which was connected to a motor. The infrared emitters and photodetectors that monitored the eye movements of a subject trying to track this ball were mounted on the special eyeglasses shown here worn by Dr. A. Terry Bahill. Two light-emitting diodes used to monitor head movements sit directly on top of the head, and the third is at the end of the stalk.*

This, like all our constraints, was designed to give our subjects the best possible chance of keeping their eyes on the ball. A low curve ball thrown by a right-handed pitcher would have been much harder to track.

By controlling the speed of the motor and counting the rotations of the shaft, we could compute the position of the ball at every instant of time, and thus compare the position of the ball to the position of the batter's gaze. We define both positions in terms of the horizontal angle of the ball: the angle between the line of sight pointing at the ball and a line perpendicular to the subject's body (see Figure 59). This angle is slightly more than 0 degree when the pitcher releases the ball, and it increases to 90 degrees when the ball crosses the plate.

We monitored horizontal eye movements with a photoelectric system using infrared emitters and photodetectors mounted on spectacle frames (as shown in Figure 58) and aimed at the sides of the blue or brown irises. As the eye moved horizontally, the amount of reflected infrared light changed, causing a variation in the current of the photodetectors. Amplifying the difference in the currents of the two detectors produced a voltage proportional to horizontal eye position. Although we sampled each millisecond, our data were filtered and compressed to produce position traces with a bandwidth of 30 Hz and velocity traces with a bandwidth of 7 Hz.[4] Since we deliberately configured the experiment to minimize vertical target

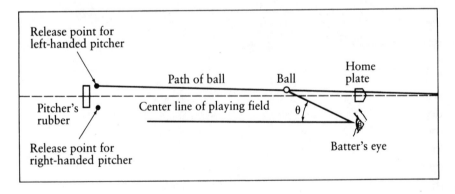

Figure 59   *The horizontal angle of the ball, θ, as defined in this study, ranges from near 0 degrees when the pitcher releases the ball, to 90 degrees when the ball crosses the plate. (From A. Terry Bahill and T. LaRitz, American Scientist, 72, 1984, pp. 249–253.)*

movements, vertical eye movements, which were measured with electro-oculography but are not shown in our figures, were negligibly small.

Head movements were monitored with a TV camera mounted on the ceiling, looking down on the subject's head. Two light-emitting diodes (LEDs) were placed on top of the subject's head, and a third LED was mounted on a stalk 7.8 inches above the head, as shown in Figure 58. The camera's signal was digitized and stored in a computer. The coordinates of the centers of the three LEDs were computed, and from these, we computed the yaw, pitch and roll angles, as well as the lateral and forward-backward translations of the head.

Yaw is rotation about the vertical axis, as occurs when shaking your head when answering no to a question. Pitch is rotation about a horizontal axis (specifically the y axis of Figure 10 in Chapter 2), as occurs when shaking your head when answering yes to a question. When a ship alternately dips its bow and stern, it is said to be pitching. Roll is rotation about a different horizontal axis (specifically the x axis of Figure 9 in Chapter 2), as occurs when cocking your head. When a ship rocks from side to side, it is said to be rolling.

We ran several subjects in our experiment, including graduate students, undergraduate students on the university baseball team, and Brian Harper, then a member of the Pittsburgh Pirates; all had 20/20 uncorrected vision.

Gaze Direction While Tracking the Ball

Figure 60 shows the data of one experiment, which were typical of the results obtained with students. This subject tracked the ball well (less than 2 degrees error) until the ball was 9 ft in front of the plate, then he fell behind. When the ball was 2 ft in front of the plate, the image of the ball was 34 degrees off his fovea. To see what it means for an object to be 34 degrees off your fovea, hold a 1-foot ruler at arm's length. If you look at the left end, the right end will be about 34 degrees off your fovea. While looking at the left end, you will not be able to read the numbers on the right end. The ball covered the distance between 9 ft and 2 ft in front of the plate in 67 milliseconds (ms) for an average angular velocity of more than 500 degrees per second—much too fast for humans to track. The maximum smooth pursuit velocity occurred just before the ball crossed the plate: the eye was going 50 degrees per second, and the head was going 20 degrees per second, giving a gaze velocity of 70 degrees per

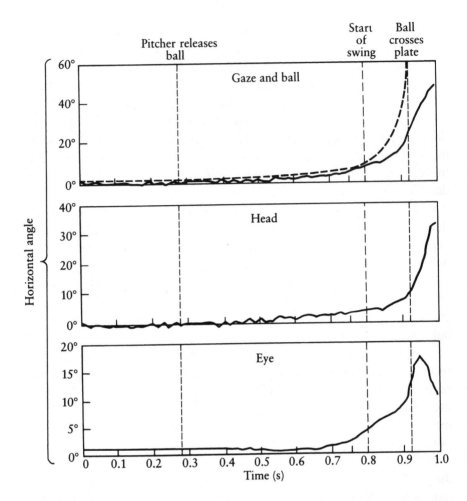

Figure 60   The solid line in the top graph represents the horizontal angle of our experimental 70-mph pitch as it would be seen by a right-handed batter facing a left-handed pitcher; the dashed line represents the horizontal angle of gaze of the subject, a graduate student, trying to track this ball. This gaze graph is generated by combining horizontal head angle (middle graph) and horizontal eye angle (bottom graph). Movements to the right appear as upward deflections. "Start of swing" is defined as the time when the horizontal velocity of the bat toward the ball became greater than zero. (From A. Terry Bahill and T. LaRitz, American Scientist, 72, 1984, pp. 249–253.)

second. This subject used primarily eye movements to track the ball; the head movements came too late in the trajectory to be of use in tracking. Other students used only head movements to track the ball. After the ball crossed the plate, this subject (like most of our students) made large head and eye movements.

Figure 61 shows the results produced by the professional ballplayer Brian Harper while tracking three of our pitches. Let us look in detail at the second pitch, Figure 61b. He tracked the ball using head and eye movements, keeping his eye on the ball longer than our other subjects did. He was able to keep his position error below 2 degrees until the ball was 5.5 ft from the plate, at which point he began to fall behind. When the ball was 2 ft in front of the plate the image of the ball was already 16 degrees off his fovea and was going 1,100 degrees per second. Thus, he was no longer tracking the ball.

While tracking the ball, as shown in Figure 61b, the eye angle changed by 17 degrees and the head horizontal rotation angle (yaw) changed by 13 degrees. Thus, he used his head and his eyes to track the ball.*

The peak velocity of his smooth-pursuit tracking was 120 degrees per second; at this point his head velocity was 30 degrees per second, thus producing a gaze velocity of 150 degrees per second. In three experimental pitches to the professional athlete, at speeds of 60, 67 and 70 mph, the overall tracking patterns were the same; his maximum smooth-pursuit eye velocities were 120, 130 and 120 degrees per second, respectively. These numbers are much faster than those commonly reported in the oculomotor literature. The fastest smooth pursuit eye movements reported in the literature are only 90 degrees per second. Therefore, this professional baseball player has truly remarkable eye tracking capability.*

Our computer algorithms for producing the gaze graphs of the professional athlete (Figure 61) were more complicated than those used for the gaze graph of the student in Figure 60 in that, in addition to combining eye angle and head angle, they also take into account the side-to-side and front-to-back movements of the head; such translations of the head

---

*Because the movement of the ball was primarily horizontal, the measured head pitch and roll angles changed very little, that is, by less than 2 degrees.

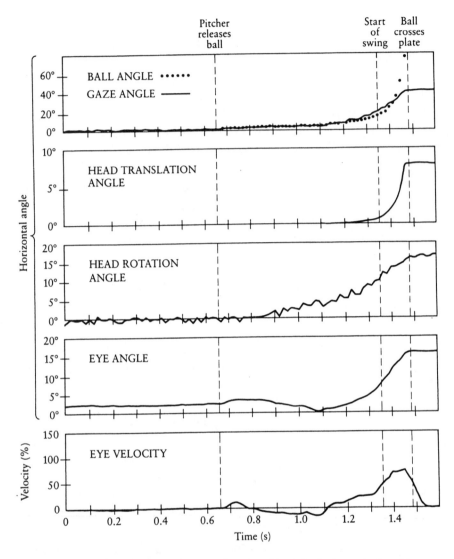

Figure 61a *The success of a professional baseball player in tracking our experimental 60-mph pitch is shown in graphs, with the same format as in Figure 60, except that the horizontal-gaze angle also takes into account the head-translation angle shown in the middle graph, which represents the eye movements necessary to compensate for side-to-side and front-to-back movements of the head.*

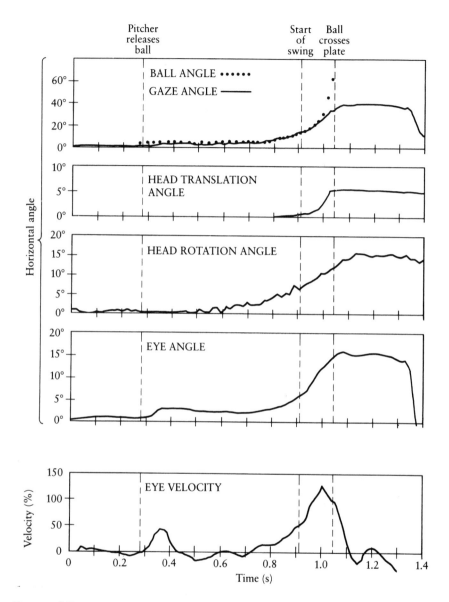

Figure 61b   The second pitch to our professional athlete, a 67-mph pitch. Same format as in Figure 61a. (Adapted from A. Terry Bahill and T. LaRitz, American Scientist, 72, 1984, pp. 249–253.)

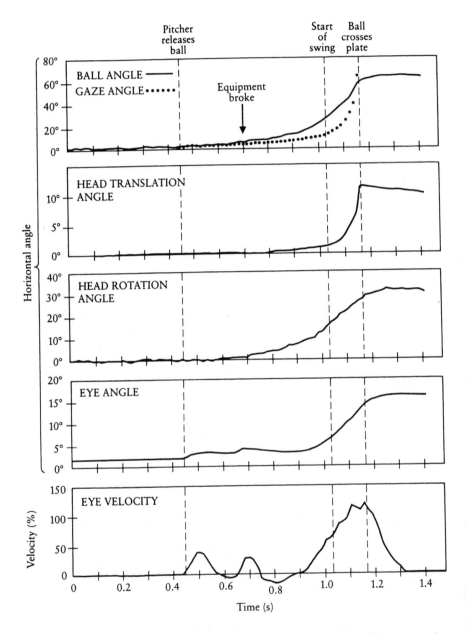

Figure 61c   *The third pitch to our professional athlete, a 70-mph pitch. Same format as in Figure 61b. The equipment broke at the point indicated by the arrow and the ball began to slow down. However, the subject continued to track as if nothing had happened. This caused his eye to get ahead of the ball.*

can produce changes in the gaze angle.[5] The data show that the contribution of this translation angle was slight until the ball was almost over the plate.

We have often heard the batting axiom "Don't move your head." So we thought the vestibulo-ocular system would not be used when tracking a baseball. However, in monitoring the eyes of the professional ballplayer, we detected a small vestibulo-ocular eye movement to the left during the early part of the ball's trajectory, as the head was moving to the right; this appears as the downward deflection between 0.8 and 1.1 s in the eye trace of Figure 61a. At this point, the head position was changing faster than the angular position of the ball, and the vestibulo-ocular eye movement compensated for the premature head movement. The batter was apparently giving his head a head start. Why would he want to do this? Because the head is heavier than the eye and consequently takes longer to get moving; therefore, in the beginning of the movement, as the head starts turning to the right (ahead of the ball), the vestibular system in the inner ear signals the oculomotor system to make a compensating leftward eye movement. This allows the head to start turning while the gaze remains fixed on the ball.

However, this vestibulo-ocular compensation must soon stop. In the last half of the ball's flight, the eye and head must both be moving to the right, and the batter must, therefore, suppress his vestibulo-ocular reflex, so that the tracking head movement does not produce compensating eye movements that would take his eye off the ball. The professional athlete was very good at suppressing his vestibulo-ocular reflex. Most of our student subjects tracked the ball with either head movements alone or eye movements alone, but not both. Perhaps they did this because they could not suppress their vestibulo-ocular reflexes as well.

There is one other noteworthy difference between the tracking of the students and the professional athlete. After the ball crossed the plate, the students usually made large eye or head movements, whereas the gaze of the professional athlete was quite steady.

The stance of our professional athlete was very repeatable. At the beginning of the pitch, his head position was the same (within 1 degree) for each of the three experimental pitches we recorded. When he was looking at the ball in the beginning of the experiment, his eyes were rotated 22 degrees to the left; his head was rotated left 65 degrees (yaw), was bowed down 23 degrees (pitch) and was tilted right 12 degrees (roll).

Obviously, the professional athlete had faster smooth-pursuit eye movements than our other subjects did. In fact, he had faster smooth-pursuit eye movements than any reported in the literature. He also had better head-eye coordination, as shown in the region where he was giving his head a head start. In addition, he was better at suppressing the vestibulo-ocular reflex and making his head and eyes move in the same direction. This allowed him to track the ball with equal-sized head and eye movements, whereas the other subjects usually tracked with either head or eye movements. Together these three superiorities allowed the professional athlete to track the ball to 5.5 ft in front of the plate compared to our students who fell behind at the 9-ft mark.

Limitations

We caution that our study, although it does warrant some useful generalizations, has limitations. First, our data pool was small. These experiments were difficult to perform, and from our 50 hours of data collection (and thousands of hours setup time and data analysis) we have complete head and eye data for 6 pitches, and partial data for another 15—a total of 11 seconds of data. Second, because we were afraid that they would break our equipment, we never let our subjects actually swing the bat at the ball; it is possible that head-eye coordination would be different if the subjects did swing the bat. Third, we produced the easiest pitch for the batter to track: a high-and-outside fastball thrown by an opposite-handed pitcher, in this case a left-handed pitcher to a right-handed batter. Nevertheless, even with these limitations, we can reasonably generalize.

No One Can Keep Their Eyes on the Ball

Although the professional athlete was better than the students at tracking our fastball, it is clear from our experiments that batters, even professional batters, cannot keep their eyes on the ball. Our professional athlete was able to track the ball until it was 5.5 ft in front of the plate. This could hardly be improved on; we hypothesize that the best imaginable athlete could not track the ball closer than 5 ft from the plate, at which point it is moving three times faster than the fastest human could track.

When questioned about our data, several professional baseball players have said things like "I try to watch the ball all the way to the plate, and if I'm hitting well, I do think I can follow it all the way."[6] When asked what the ball looked like at the time of collision, they said "like a ball." When shown the picture of Figure 35 in Chapter 5, they usually responded with "Well, maybe I don't really see the ball hit the bat."

AS SMART AS HE WAS, ALBERT EINSTEIN COULD NOT FIGURE OUT HOW TO HANDLE THOSE TRICKY BOUNCES AT THIRD BASE.

One question we are often asked: "How slow would the ball have to go, so that the batter could keep his eye on the ball?" This question, however, seems to relate to linear velocity, whereas it is the angular velocity that is important. (Angular velocity is the rate of change of the angle θ in Figure 59.) The ball's linear velocity as it travels between the pitcher's mound and the plate decreases by 10 percent. But its angular velocity with respect to the batter changes drastically. To illustrate this point, note that if you are standing on the ground, it is easy to track for minutes an airplane flying at an altitude of 20,000 ft. but it is not possible to track, for even a few seconds, a jet plane flying 100 feet directly overhead. In both cases, the airplane may have the same linear velocity, but the angular velocities are drastically different. So the head and eye movements needed to track it are much different. To help answer the question of how slow the ball would have to go to allow tracking, we ran a computer simulation of the batter tracking the ball. Figure 62 shows the results of this simulation. It shows the distance in front of the plate where the eyes would fall behind the ball for different pitch speeds. We were very conservative in determining this distance. We let the computer track up to 300 degrees per second; which is twice as fast as our professional athlete and three times as fast as any human reported in the oculomotor literature over the past 50 years. No data points are shown at or below 20 mph, because such a pitch would hit the ground before it crossed the plate. According to this equation from Chapter 2

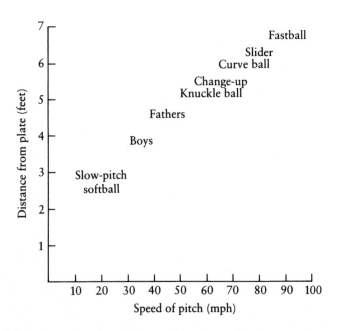

Figure 62   How close can you track the ball? Distance from the plate when the best imaginable athlete could no longer keep his eye on the ball is plotted against the speed of the pitch.

$$R = \frac{2v_{x0}v_{z0}}{g}$$

the slowest pitch that could reach the plate would be 21 mph,* assuming no air resistance, and that $v_{x0} \geq v_{z0}$; that is, it was thrown at an angle of 45 degrees or less. (This assumption is reasonable, because if $v_{x0} < v_{z0}$ then you would just lose the ball in the vertical plane before the horizontal plane.) We have also labeled Figure 62 with speed information taken from Table 1 in Chapter 3. It shows, for example, that seven-to-nine-year-old boys throw the ball at speeds between 30 and 40 mph. A batter trying to hit a pitch thrown by such a boy would fall behind when the ball was 3 to 4 ft in front of the plate. This leads to the quite astounding conclusion that the batter can't keep his or her eye on the ball even in Little League or slow pitch softball!

The Alternate Strategy

This finding, that no athlete could track the ball within 5 ft of the plate, runs contrary to one of the most often repeated axioms of batting instructors—"Keep your eye on the ball"—and makes it difficult to account for the widely reported claim that Ted Williams could see the ball hit his bat.

If Ted Williams were indeed able to do this, it could only be possible if he made an anticipatory saccade that put his eye ahead of the ball and then let the ball catch up to his eye. This was the strategy employed by the subject of Figure 63; this batter observed the ball over the first third of its trajectory, predicted where it would be when it crossed the plate, and then made an anticipatory saccade that put his eye ahead of the ball, thereby allowing the ball to catch up to his eye. Using this strategy, the batter could see the ball hit the bat. However, let us note that in a personnel communication Ted Williams stated that it was "only on very rare occasions" that he could see the ball hit the bat. It seems that he rarely used this strategy of taking his eye off of the ball, so that he could see it hit his bat.

If a batter does use this strategy, the anticipatory saccade must be made before the end of the ball's trajectory, because saccadic suppres-

---

*For the shorter 46-ft distance between the rubber and home plate of Little League and softball, the slowest pitch that could reach the plate would be 18 mph.

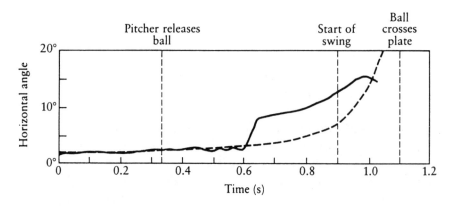

Figure 63 *In order to see the ball hit his bat, this batter made an anticipatory saccade, indicated by the jump in the gaze angle (solid line) at 0.6 s, that put his eye ahead of the ball (dashed line); as a result, the ball was on the fovea at the point of contact. The subject did not move his head until after the ball crossed the plate. (From A. Terry Bahill and T. LaRitz, American Scientist, 72, 1984, pp. 249–253.)*

sion prevents us from seeing during saccades.[7,8] To see the effects of this suppression look at the image of your right eye in a mirror. Now look at the image of your left eye. Did you see your eyes move? Probably not, because we have a process that turns off the visual system during saccadic eye movements. This is a comforting thing to do, because otherwise every time we made a saccadic eye movement, we would think that the world was flying around us. To prevent such confusion we suppress vision during saccades. This suppression of vision exists for about 20 ms before and after the saccade. So if you want to see the ball hit the bat, you must make your anticipatory saccade early in the trajectory. This means that you have to take your eye off the ball at precisely the time when you want to be seeing the ball the best. Using an anticipatory saccade to put your eye ahead of the ball may be an oft-trained strategy, but it is probably not the best strategy for hitting the ball.

On the other hand, why would a batter want to see the ball hit the bat? He could not, because of his slow reaction time, use the information gained in the last portion of the ball's flight to alter the course of the bat. We suggest that he uses the information to discover the ball's actual trajectory; that is, he uses it to learn how to predict the ball's location when it crosses the plate—how to be a better hitter in the future. The batter says to himself, "I thought the ball was going to be right there. But instead it was a half-inch lower. I will predict better the next time the pitcher throws that pitch." Figure 61c offers further evidence that the batter uses his predictive abilities to help track the ball. At the point indicated by the arrow, our string broke and the ball began to decelerate. But the batter ignored the real world and continued to follow his predictions, which put his eye ahead of the ball.

We have shown that no one could keep his eye continuously on the ball as it flies from the pitcher to the plate. For our professional athlete, the ball was always more than 2 degrees off his fovea before it came within 5 ft of the plate. However, when the ball is off your fovea, you can still see with peripheral vision. However, with peripheral vision, the ball would only appear as a white blur; you would not see details. For example, with peripheral vision you could not count the red stitches on the ball, read the label on the bat or see whether the ball hit the bat in the center or an eighth of an inch below. So we remain confident in our statement that batters cannot keep their eyes on the ball.

It has often been said that athletes are dumb. Our studies have shown the contrary. They are not paid a million dollars for their six-month job because of their bodies; it is because of their brains. The players that are paid the most have the best brains: they can predict the flight of the pitch better than other mere mortals can.

Other Movement Systems

Batters do not use vergence eye movements. This is reasonable, since negligibly small vergence eye movements are needed to track the ball between 60 and 6 ft from the plate, and since there is not sufficient time to make such movements between 6 ft and the point of contact; indeed, our data contained no vergence eye movements. So any claim that a batter actually saw the ball hit the bat must be based on monocular vision; only one eye tracks the ball.

The fact that our professional athlete used his head to help track the ball seems to violate another often-repeated batting axiom: "Don't move your head." The professional made tracking head movements between 10 and 20 degrees, which were probably small enough to go unnoticed by a coach. What the batter does not want to do is to allow rotations of the body to drag the head along. These body movements could produce head movements of 90 degrees or more, which, along with correlated poor performance, would be noticed by a coach. Therefore, we think the axiom should be protracted: "Don't let your body pull your head, but it's okay to move your head to help track the ball."

If we can change the subject for a moment from the batter to the umpire, we find some interesting advice in *The Umpire's Handbook.*[9] It says, "The umpire should follow the ball with his eyes only." We do not think the authors really meant this, because they follow with phrases like "If he moves his body," and "jerking your head around." I think they meant that it is detrimental to tracking success if body movements are allowed to induce head movements. The data of this chapter indicate that batters and umpires should track the ball normally, that is, with head and eye movements. However, the head movements must be small tracking movements that will not be seen by a faraway observer, not large body-induced head movements.

Does It Curve or Drop?

In Chapter 4 we stated that the curve ball can curve horizontally and drop vertically. We explained the advantage of the drop by noting that the sweet spot on the bat is about 6 inches long and only half an inch high. Thus, a drop would be much more effective at taking the ball away from the bat's sweet spot than a curve. However, we have observed a Randy Johnson curve ball that dropped 2 ft but also curved 2 ft. Why would he want it to curve?

It is well known in baseball that right-handed batters have significantly greater success hitting against left-handed pitchers than against right-handed pitchers. The obverse is true for left-handed batters. This difference in most batters' ability is perplexing, since our computer analysis shows that an overhand fastball thrown by a right-handed pitcher to a right-handed batter would have started with an initial horizontal angle of +1 degree instead of the +3 degrees for a left-handed pitcher; this 2-degree offset in the beginning is completely insignificant. Furthermore, over the last third of the pitch, there is no difference in either angular position or angular velocity. We suspect, with little evidence, that for a right-handed batter, the right-handed pitcher is harder to hit than the left-handed pitcher, only for the curve ball, not for the fastball.

The advantage of the opposite-handed batter is conjectured by Karl Schurr of Bowling Green State University to be nothing more than fear:

When a right-handed batter sees a ball thrown by a right-handed pitcher coming straight at his head he doesn't know if it is a curve ball that will cross the plate for a strike, or a wild fastball that will hit his head. On the other hand, the left-handed pitcher would have to be wild as a March hare to raise as much adrenaline in the batter.

We conclude therefore that the curve ball should drop as much as possible in order to avoid the bat, but it should also curve to induce fear and confound the batter. We would expect right-handed pitchers to make their curve balls primarily drop when throwing to left-handed batters, and to substitute more horizontal curve when throwing to right-handed batters.

The Perceptual Illusion of the Rising Fastball

Over the last century, many baseball players have described the *rising fastball*: this pitch starts normally, but right in front of home plate, the ball jumps up a half-foot, making it hop over the bat. Such behavior is impossible according to the laws of physics. Our simulation and model, based on principles of physics, psychology and physiology, explain this contradiction. Although the numbers given in this section are for professional baseball players, the simulation and model apply to all players, right down to Little Leaguers. They can also be extended to cricket.[10] This section on the rising fastball is based on A. Terry Bahill and William J. Karnavas (1993).[11]

The rising fastball could be defined as a pitch where the ball (1) seems to jump up, right in front of the plate, (2) crosses the plate above the pitcher's release point, (3) is going upward when it crosses the plate or (4) falls less than would be expected due to gravity. By definitions (2) and (3), a rising fastball could be thrown by a sidearm baseball pitcher or a softball pitcher, but not by an overhand baseball pitcher. For example, an overhand pitch is released about 6 feet above the ground; if the ball crossed the plate higher than this it would not be a strike. And in order for the ball to be going upward when it crosses the plate, it would have to fall initially and then, near the end of its flight, experience an upward force that is greater than that of gravity. A force that opposes gravity is produced by the backspin on a fastball. However, the maximum spin rate ever measured for a human pitch, -2300 revolutions per minute (rpm), would only create a force two thirds that of gravity. So, although a fastball's lift due to backspin may not overcome gravity, it does make it fall less than would be expected due to gravity, which is definition (4). Therefore, all fastballs fit this definition, making it trivially simple and uninteresting. Thus, for the rest of this section we will only consider overhand pitches in baseball and definition (1). For more discussion about batters' perceptions of the rising fastball see Michael McBeath,[12] who has summarized the comments of baseball players and scientists over the last three centuries.

The Simulation

Figure 64a and Table 12 show our simulations of 95 and 90 mile per hour (mph) fastballs. In these simulations, both pitches were launched one degree upward with 1500 rpm of backspin. The distance between the front of the pitcher's rubber and the tip of the plate is 60.5 feet. But the pitcher releases the ball about 5 feet in front of the rubber. Therefore, for these simulations, the pitcher's release point was 55.5 feet away from the tip of the plate. The bat hits the ball about 1.5 feet in front of the batter's head, which was assumed to be aligned with the front of the plate. The plate measures 17 inches from the tip to the front edge. So in these simulations the point of bat-ball collision was 3 feet in front of the tip of the plate, which is represented in the bottom rows of Table 12. The pitcher's release point was assumed to be 6 feet high. Later, we will show how sensitive our conclusions are to these numbers. The numbers in Table 12 are more accurate than those in Figure 64, because they were made with a more sophisticated model.

Figure 64   (a) Computer simulation of the trajectory of a 95-mph fastball (solid line and circles) and a 90-mph fastball (dashed line and triangles). The slower pitch takes longer to get to the plate and therefore drops more. (b) Computer simulation of the trajectory of a 95-mph fastball (solid line and circles), and the batter's mental model of this trajectory (dashed line and triangles) when he underestimated the speed of the pitch by 5 mph. (c) The same simulation as Figure 64b, except that when the ball was 20 feet in front of the plate, the "batter" realized his mental model was wrong and corrected it, thus putting his mental model triangles on the 95 mph trajectory. (From A. Terry Bahill and W. J. Karnavas, Journal of Experimental Psychology: Human Performance and Perception, vol. 19, no. 1, 1993, pp. 3–14.)

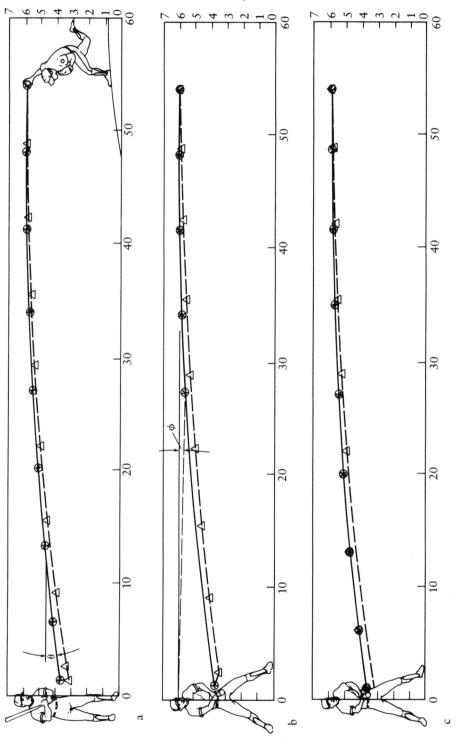

Table 12 Trajectories of Fastballs

	95 mph fastball			90 mph fastball		
Time Since Release (msc)	Distance (ft)	Height (ft)	Speed (mph)	Distance (ft)	Height (ft)	Speed (mph)
0	55.5	6.00	95.0	55.5	6.00	90.0
50	48.6	5.86	93.3	49.0	5.86	88.5
100	41.8	5.67	91.7	42.5	5.68	87.0
150	35.2	5.43	90.2	36.2	5.44	85.6
200	28.6	5.15	88.6	30.0	5.16	84.3
250	22.2	4.83	87.2	23.8	4.84	82.9
300	15.8	4.46	85.7	17.8	4.47	81.6
350	9.6	4.05	84.4	11.9	4.05	80.4
400	3.4	3.59	83.1	6.0	3.59	79.2
404	3.0	**3.56**	83.0			
426				3.0	**3.33**	78.6

The Qualitative Model for the Rising Fastball

The illusion of the rising fastball could be the result of the batter under-estimating the speed of the pitch. For example, suppose the pitcher threw a 95-mph fastball, but the batter underestimated its speed and thought it was only going 90 mph. He would expect to hit the ball 3.33 feet above the ground (from the sixth column, 426-ms row of Table 12). But if he were to take his eye off the ball (indicated by the absence of some circles in the actual 95 mph pitch of Figure 64b) and look at his bat in order to see the expected bat-ball collision, then the next time he saw the ball it would be 3.56 feet above the ground (third column, 404 ms row of Table 12), 3 inches above his bat. He might explain this by saying, "The ball jumped 3 inches right in front of the plate." Should the batter underestimate the speed at 80 mph, he would say the ball jumped 1 foot. But the manager in the dugout, the catcher and the center field TV camera, having different perspectives, would not see this jump.

Batters use one of two strategies in tracking the pitch.[1] *The optimal learning strategy*, which allows the batter to see the ball hit the bat, is track the ball over the first part of its trajectory with smooth-pursuit eye movements, make a saccadic eye movement to a predicted point of bat-ball collision, continue to follow the ball with peripheral vision, letting the ball catch up to the eye and, finally, at the end of the ball's flight, resume smooth-pursuit tracking with the images of the ball and bat on the fovea. It is called the optimal learning strategy, because the batter observes the ball, makes a prediction of where it will cross the plate, sees the actual position of the ball when it crosses the plate and uses this feedback to learn to predict better the next time the pitcher throws that pitch. The *optimal hitting strategy*, which does not allow the batter to see the ball hit the bat, is track the ball with smooth-pursuit eye movements and fall behind in the last 5 feet. It is called the optimal hitting strategy, because the batter keeps his eye on the ball longer, which should allow him to make a more accurate prediction of where the ball will cross the plate. However, we have no evidence that batters voluntarily switch between these two strategies.

With the optimal learning strategy a batter would perceive more rising fastballs, because his eye would not be on the ball (which is why we removed some circles from Figure 64b) when he started his swing, which occurs 100 to 150 ms before bat-ball contact when the ball is about 20 feet in front of the plate, and therefore during the pitch the batter could not discover inaccuracies in his estimation of pitch speed and make adjustments. Whereas with the optimal hitting strategy a batter would perceive fewer rising fastballs, because his eye would be on the ball when it was 15 to 25 feet from the plate, therefore he could sense inaccuracies in his speed estimation and, as shown in the last 150 ms of Figure 64c, make an appropriate correction, such as checking his swing.

The batter does a different thing in each third of the pitch's flight: observe, compute and swing. For the rising fastball the batter observes the ball for the first third of its flight, he underestimates the pitch speed and therefore miscomputes the height of the bat-ball collision point, then at the start of his swing he takes his eye off the ball to look at the estimated bat-ball collision point, and when the ball comes back onto his fovea it is higher than he thought it would be.

In the next section, we present a quantitative model to investigate the perceptual illusion of the rising fastball. This model for human brain systems is based on D. M. Regan[10,13,14] and J. T. Todd[15] and uses only experimentally verified visual processes.

The Quantitative Model for the Rising Fastball

The batter must predict precisely where the ball will be in three-dimensional space at some specific future instant. This judgment involves four coordinates: the x, y and z spatial coordinates shown in Figure 65 and t, the time coordinate.

It is important to note that the batter has no direct visual access to the x, y and z spatial coordinates: his judgments must be based entirely on retinal image data. The relevant retinal parameters are the angular size of the ball, γ, and the angular distance of the ball's image off the fovea, Φ, shown in Figure 65, and their time derivatives dy/dt and $d\phi/dt$. We assume the batter's eyes are in the x-y plane as shown in Figure 65: although the Pittsburgh Pirate studied by Bahill and LaRitz[1] actually rotated his head 23° in pitch and 12° in roll.

To hit the ball the batter must predict *when* and *where* it will cross the plate. First, let us see how the batter can judge *when*. In his novel *The Black Cloud*, Sir Fred Hoyle[16] showed that the time-to-contact (τ) with an object moving along the line-of-sight can be approximated with

$$\tau = \frac{\gamma}{d\gamma/dt}$$

where γ and dy/dt are respectively the object's angular size and rate of change of angular size. It has been suggested that birds use time-to-contact when diving into the water to catch prey, and athletes use time-to-contact when jumping to hit a dropped ball, adjusting strides when running hurdles and timing their swings in table tennis; for these tasks time-to-contact is judged with an accuracy around 2 to 10 ms. Cricket players time their swings with an accuracy of ±5 ms.[10] The batter's calculation of time-to-contact has three sources of error. First, the above equation is only an approximation, because it uses the approximation Tan $\psi = \psi$. Second, the ball is not headed directly at the batter's eye. In our simulations, these two sources produced errors of less than one ms until after the swing had begun. The third source of error, which results from the batter hitting the ball 1.5 feet in front of his eyes, produces a constant 11 ms of error.

The human visual system can implement the equation

$$\tau = \frac{\gamma}{d\gamma/dt}$$

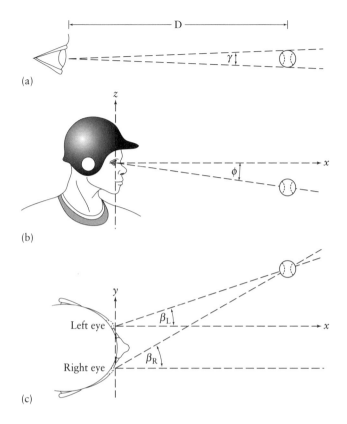

Figure 65   *Visual system parameters used by the batter: (a) angular size of the baseball, γ, the batter has no means to sense the distance to the ball (D); (b) distance of the ball's image off the batter's fovea, Φ; and (c) horizontal angle of the right and left eyes, $β_R$ and $β_L$. (From A.Terry Bahill and W. J. Karnavas,* Journal of Experimental Psychology: Human Performance and Perception, *vol. 19, no. 1, 1993, pp. 3–14.)*

First, there is psychophysical evidence that the human brain contains units tuned to size (γ), and size-tuned neurons have been found in monkey visual cortex. Second, psychological studies have shown that the visual system has specialized "looming detectors" that compute dy/dt independent of the object's trajectory. Furthermore, specific brain neurons are sensitive to changing-size dy/dt. Using these two pools of neurons, the brain could compute.

$$\tau = \frac{\gamma}{d\gamma/dt}$$

We conclude that from the time the ball leaves the pitcher's hand, the batter's retinal image contains accurate cues for time-to-contact, and that the human visual system is capable of using these cues. Evidentially most batters estimate when the ball crosses the plate with an accuracy of at least ± 9 ms, otherwise the ball would be hit foul. Next, we want to consider the more difficult issue of judging *where* the ball will be at the time of bat-ball contact.

The batter can estimate the ball's height at contact from the ball's vertical speed and the time since the pitcher released the ball. He can estimate the ball's vertical speed from the retinal velocity and the distance to the ball. The batter has an accurate sense of retinal velocity $d\phi/dt$. However, and this is the crucial element in our explanation of the illusion of the rising fastball, retinal image information provides poor cues to the absolute distance to the ball (D in Figure 65). Classical stereoscopic depth perception is of little help in this regard: although the stereoscopic depth system provides a precise indication of relative depth (i.e., the difference between the x-axis distances of two objects imaged near the fovea), it provides little indication of absolute distance. In tracking the pitched ball, the batter has one object, the ball, imaged on his fovea. Therefore, the batter cannot measure distance to the ball; he can only estimate it.

Our psychophysical explanation for the rising fastball is as follows. The batter can estimate pitch speed and the time since the ball left the pitcher's hand. He can use this data in conjunction with his experience to estimate the distance to the ball ($\hat{D}$). The "hat" on top of a symbol indicates that it is an estimate. He can then use this estimate for distance and the ball's retinal image velocity ($d\phi/dt$) to estimate the vertical velocity ($d\hat{z}/dt$). From the vertical velocity and the time-to-contact (τ) he can estimate how far the ball will fall in the last part of its flight, and therefore estimate the height of the ball when it crosses the front edge of the plate $\hat{z}(plate)$. If any of these computations are wrong, then his estimate for the ball's height at contact will be wrong. For example, suppose the pitcher enhances the batter's expectations with a series of 90-mph pitches, and then throws an unusually fast 95-mph pitch. The batter is likely to use a 90-mph mental model to interpret the retinal image information about the 95-mph pitch. Suppose the batter tried to estimate the ball's vertical speed 250 ms after the ball left the pitcher's hand. If the actual pitch were a 95-mph fastball, it would be 20.7 feet from his eye, subtracting 1.5 feet (the distance between the tip of the plate and the batter's eye) from the x distance of 22.2 in Table 12. Its vertical velocity of -8 ft/sec (from velocity $= gt$) would, at this distance, produce a retinal velocity of -21 deg/sec. However, if the batter thought the pitch was a 90-mph fastball, he would think that it was 22.3 feet away 250 ms after release. At this distance, a retinal image velocity of -21 deg/sec would indicate that the vertical velocity was about -8.6 ft/sec. So the batter would think the ball was falling faster than it really was and would probably swing under the ball. Therefore, if the batter made a saccadic eye movement to a predicted point of bat-ball contact, this point would be below the ball when the ball caught up with the eye, and the ball would seem to have jumped upward, in this example by 3 inches. To be sure, this error of visual judgment could be avoided if the batter had an accurate visual cue to the ball's absolute distance (D) or its speed, but, as we have seen, the batter is essentially "blind" to these two important parameters.

This model can be summarized with Figure 66 and the following equation.

$$\hat{z}_{from-eye}(plate) = (\hat{D}_0 - \hat{t}_{sr}\hat{s})\left(\frac{d\phi}{dt}\frac{y}{dy/dt}\right)$$

where $\hat{z}_{from-eye}(plate)$ is the estimated vertical distance from the batter's eye when the ball crosses the front edge of the plate, $\hat{D}_0$ is the estimated distance between the ball and the batter's eye at the time of release, $\hat{t}_{sr}$ is the estimated time since release, $\hat{s}$ is the estimated pitch speed, $d\phi/dt$ the retinal image velocity, γ is the retinal image size and dy/dt is the rate of change of retinal image size. The first term of the above equation is the estimated distance to the ball, $\hat{D}$, and the last term is τ. So the above equation can be rewritten as

$$\hat{z}_{from-eye}(plate) = \hat{D}\frac{d\phi}{dt}\tau$$

Overestimating any one of these three terms could produce the illusion of the rising fastball. But as we have already shown, batters can accurately compute $d\phi/dt$ and τ; the difficulty is in estimating the distance to the ball. And in this model, they use estimated pitch speed to estimate the distance to the ball. That makes the speed esti-

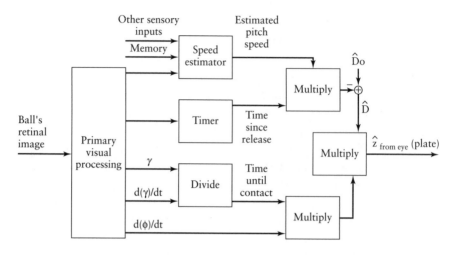

Figure 66   *A model for a batter trying to estimate the height of the ball at the time of its collision with the bat. The input to the system is the ball's retinal image and its output is the estimated height of the ball at the time of bat-ball contact. (From A. Terry Bahill and W. J. Karnavas, Journal of Experimental Psychology: Human Performance and Perception, vol. 19, no. 1, 1993, pp. 3–14.)*

mator in Figure 66 the crucial element of this model. In order to estimate the height of the ball at the time of contact, the batter must be able to estimate the pitch speed. But pitch speed cannot come from the primary visual processes. The speed estimator receives input from the primary visual processes, but this arrow is unlabeled because we do not know exactly which signals are involved. The speed estimator probably uses memory and other sensory inputs: some visual, such as the motion of the pitcher's arms and body, and some auditory, such as a whistle from a coach who might have stolen the catcher's signals.

The speed estimator might be most accurate just after the pitcher releases the ball, for at this point the batter should be able to estimate its distance $\hat{D}_0$ quite well. And the distance $\hat{D}_0$ combined with the time-to-contact would provide the speed. But at this time, the estimation of the ball's height at the time-of-contact may still be in error, because retinal image velocity $d\phi/dt$ is inaccurate until the ball is 49 feet from the tip of the plate.

We have said before that the batter must predict both *where* and *when* the bat-ball collision will occur. To hit a line drive in fair territory, the batter must estimate *when* to within ±9 ms and *where* to within $\pm\frac{1}{2}$ inch. Comparing these numbers to the 21-ms and 2.8-inch differences between the 90- and 95-mph fastballs, we see that estimating *where* requires greater percent of accuracy. And batters seem to be more accurate at estimating *when*. Indeed, few line drives are hit in foul territory, whereas there are many foul tips and pop-ups. Figure 66 suggests an explanation: *when* (τ) can be computed from primary visual processes, whereas the pitch speed, and therefore *where*, can only be estimated. So *where* is the crucial parameter: it requires greater precision, yet it cannot be calculated from primary visual processes; it can only be estimated.

Although we developed this model to explain the illusion of the rising fastball, it, of course, could also be applied to other pitches. For example, the change-up is only effective if it fools the batter and makes him *over*estimate the pitch speed. According to our model, this would make the batter swing early and over the ball. Although we have not heard coaches describe it this way, our model predicts that an effective change-up should result in a ground ball to the third baseman.

Experimental Validation of the Model

To help validate our model we ran some simple experiments using a mechanical pitching machine (the two-wheeled type). We threw 450 pitches to seven subjects: three adults and four boys aged 9, 11, 11 and 13. Nominally, the speed of the machine was set for 50 mph, but occasionally it threw a fast pitch at 55 mph. The number of normal pitches between these fast pitches was randomly chosen from among 3, 4, 5 and 6. An observer (who did not know the pattern of normal and fast pitches) recorded the relationship of the bat and ball when the ball crossed the plate. We averaged the results of the fast pitches and of the two pitches before and after, as shown in Figure 67. These data show that on the fast pitches the batters swung below the ball, which is just what would happen if they underestimated the speed of the pitch.

The pitching machine was not perfect. About 15 percent of the time it threw "balls" out of the strike zone. This randomness lessened our batters' expectations, and therefore should have lessened the effect of the rising fastball. Most of our batters "took" (did not swing at) these "balls." However, they seldom took a strike. Except, and much to our surprise, often

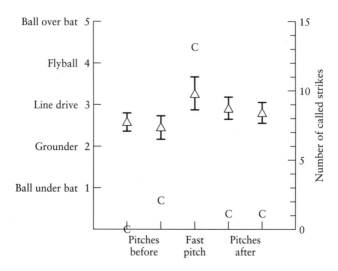

Figure 67   *Averaged data from seven batters showing that when an unusually fast pitch was thrown most batters swung under the ball or "took" a called strike. The C's represent the number of called strikes. The triangles are the mean values and the vertical bars are the 95 percent confidence intervals. (From A. Terry Bahill and W. J. Karnavas, Journal of Experimental Psychology: Human Performance and Perception, vol. 19, no. 1, 1993, pp. 3–14.)*

they took the unusually fast pitches, even when they were in the strike zone. We recorded these "called strikes" and indicate their number with the C's in Figure 67. They might have "taken" these called strikes because they underestimated the pitch speed and were therefore going to swing under the ball; but in the last third of its flight, when it would be too late to alter the swing, they realized their error, and did the only thing they could do, stop the swing. These called strikes also decreased the statistical effects of the rising fastball in our data, because our batters did not swing at the pitches that fooled them the most, the pitches that would have shown the greatest effects. The adults in our study understood the purpose of the study and swung at every pitch. It was only the boys, who did not understand the purpose of our experiments, who "took" the pitches.

The data shown in Figure 67 are the averages of all our subjects. Each of our seven individual subjects showed this same effect; but it was more emphatic in our inexperienced subjects.

So, in summary, both the randomness of our pitching machine and the batters "taking" called strikes decreased the effect of the rising fastball in our data. But in spite of these imperfections, it was still statistically significant that when the unusually fast pitch was thrown, the batters underestimated the speed and swung under the ball. This is precisely the illusion of the rising fastball.

The Sensitivity Analyses

To help validate our simulation and model we will now investigate how well our conclusions hold up under parametric sensitivity analyses. First we will show a sensitivity analysis of our simulation of the pitch (Figure 64 and Table 12), and then we will show a sensitivity analysis of our model of the human brain (Figure 66).

Each of the eight simulation parameters was varied by 5 percent. Then we ran the simulation and calculated the percentage change for the perceived jump. The ratio of these two percentages is the relative sensitivity.[17] These sensitivity values are shown in the top of Table 13.

To help understand these numbers note that from our nominal simulation of Table 12 the apparent jump of the rising fastball is 0.230 feet. When the horizontal position of the bat-ball collision point was moved back 5 percent, the amount of perceived jump increased to 0.231 feet, or by 0.5 percent, which is −0.1 (the value of the sensitivity coefficient from Table 13) times 5 percent (the perturbation size). Similarly, when the estimated pitch speed was decreased 5 percent, the amount of perceived jump increased to 0.472 feet, or by 105.4 percent, which is 21.1 times 5 percent. Therefore, the simulation is most sensitive to pitch speed and the estimated pitch speed; in comparison, all other parameters are insignificant.

It is rewarding to see that the pitch speed and the estimated pitch speed are the most important parameters of the simulation. The results of this sensitivity analysis tell us that we should not (1) try to get more accurate values for the other parameters of the simulation, (2) customize the simulation for individual players or (3) try to get more accurate values for the thresholds of the visual parameters γ or $d\phi/dt$, because these parameters are not that important. The bottom line of this sensitivity analysis is that our simulation is a good representation for the pitched baseball.

Next, we did a sensitivity analysis of our model of Figure 66 for a 90-mph pitch. Each of the six model parameters was varied by 5 percent,

Table *13* Results of Sensitivity Analyses

Sensitivity Coefficients for the Simulation

Parameter	Value
Horizontal position of the pitcher's release point	−0.2
Vertical position of the pitcher's release point	0.0
Release angle	0.0
Horizontal position of the batter's eye	0.0
Vertical position of the batter's eye	0.0
Horizontal position of the bat-ball collision	−0.1
Pitch speed	18.9
Estimated pitch speed	−21.1

Sensitivity Coefficients for the Model

Parameter	Value
$\hat{D}_0$ the distance to release point	−2.57
$\hat{t}_{sr}$ the time since release	1.57
$\hat{s}$ the estimated pitch speed	1.57
$d\phi/dt$ the retinal image velocity	−1.00
γ the retinal image size	−1.00
$d\phi/dt$ the rate of change of retinal image size	0.95

and we calculated the percentage change in how far the ball dropped between 250 ms after release and the time of bat-ball contact. The ratio of these two percentages is the relative sensitivity. These sensitivity values are shown in the bottom of Table 13.

The model is most sensitive to the three parameters that are used to estimate the distance to the ball. The batter cannot get these parameters from the primary visual processes. It seems that the model is most sensitive to the things that are most difficult for the human to compute. Perhaps superior athletes are superior not because they have superior primary visual processes, but because of their subsequent processing of this information. In other words, it is their brains not their bodies that make them superior.

Loose Ends

From these studies, we can see that estimating pitch speed is important for the batter. It seems that a useful training technique would be to use a radar gun during batting practice and announce the speed immediately after every pitch. This would help the batter learn to estimate pitch speed better.

The Breaking Curve Ball

To further test the model, we used it in a situation for which it was not designed. We asked what would happen if the batter *overestimated* the speed of the pitch? According to our model, he would perceive the ball dropping more than expected in the last few feet. The ball would break so fast it would look "like it rolled off a table." Such fast-breaking curves have been described by professional batters. However, the laws of physics say that the curve ball must exhibit a continuous curve and not an abrupt break near the plate. We will use our model to help explain the illusion of the breaking curve ball, but first we must explain the simple curve ball.

Why Does the Curve Ball Curve?

There is no longer a controversy about whether or not a curve ball curves; it does. The curve ball obeys the laws of physics. These laws say that the spin of the ball causes the curve. Should this spin be horizontal (as on a toy top), the ball curves horizontally. If it is topspin, the ball drops more than it would due to gravity alone. If it is somewhere in between, the ball both curves and drops. In baseball, most curve balls curve horizontally and drop vertically. The advantage of the drop is that the sweet spot on the bat is about 6 inches long but only one half inch high. Thus, a vertical drop would be more effective at taking the ball away from the bat's sweet spot than a horizontal curve. We now want to present the principles of physics that explain why the curve ball curves.

Our first explanation invokes Bernoulli's Principle. When a spinning ball is placed in moving air, as shown in Figure 68, the movement of the surface of the ball and a thin layer of air that "sticks" to it slows down the air flowing over the top of the ball and speeds up the air flowing underneath the ball. Now, according to Bernoulli's Equation, the point with

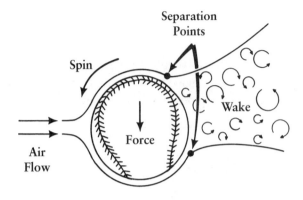

Figure 68   Air flows smoothly around the ball until it gets to the separation points. There the airflow changes into a turbulent wake that is deflected upward in this figure. If this were a top view, it would explain the curve of a ball. If it were a side view, it would explain the drop of a ball. (From W. L. Chapman, A. Terry Bahill and A. W. Wymore, Engineering Modeling and Design, CRC Press, Boca Raton, Fla., 1992.)

lower speed (the top) has higher pressure, and the point with higher speed (the bottom) has lower pressure. This difference in pressure pushes the ball downward.

Our second (and probably better) explanation involves the principal of conservation of momentum. As shown in Figure 68 the wake of turbulent air behind the ball is deflected upward. You can see such wakes on the downstream side of bridge abutments and behind boats. In a boat, swinging the rudder to the right deflects water to the right, and to conserve momentum the back of the boat must be pushed to the left. You can feel this force if you put your hand out the window of a moving car. (Make sure the driver knows you are doing this!) Tilt your hand so that the wind hits the palm of your hand at an angle. This deflects the air downward, which causes your hand to be pushed upward. Tilt your hand so that the wind hits the back of your hand. This deflects the air upward, which causes your hand to be pushed downward. Now let us relate this to the spinning baseball in Figure 68. Before the ball interacts with the air, all the momentum is horizontal. Afterward the air in the wake has upward momentum. The principle of conservation of momentum therefore requires that the ball have downward momentum. Therefore, the ball will curve downward.

There are several ways to shift the wake of turbulent air behind a baseball. The wake is shifted by the spin on a curve ball. The friction that slows down the flow of air over the top of the ball causes the air to separate from the ball sooner on the top than on the bottom, as shown in Figure 68. This shifts the wake upward, thus pushing the ball downward. For nonspinning pitches, such as the knuckleball and the scuff ball, when the seams or the scuff are near the bottom separation point, they create turbulence, which delays the separation, as shown in the bottom of Figure 68, again shifting the wake upward and pushing the ball downward.

So, when the pitcher puts spin on the ball, the wake of turbulent air behind the ball is moved to one side, causing the ball to curve, thereby confounding the poor batter who is trying to hit it.

Return to the Breaking Curve Ball

Baseballs (except for knuckleballs) follow smooth parabolic trajectories. The 90-mph fastball of Table 12 and Figure 64 falls 2.5 feet due to gravity. A plot of this vertical distance as a function of time would be parabolic ($\frac{1}{2} at^2$). The ball falls 15 inches from 250 to 400 ms, but only 10 inches in the 150 ms before that. The ball drops more in the last 150 ms than in the period right before it, but it follows a smooth parabolic trajectory. Now let us see how the ball drops with the addition of a vertical force due to spin on the ball. Table 14 shows the results of simulations of 80- and 75-mph curve balls. Both were launched at an angle of 2.5 degrees up with 1900 rpm of pure topspin. We used the formula on pages 160–161 for the lift and drag forces. Now let us look at the 80-mph pitch. We can see that between 300 and 450 ms the ball falls 21 inches, but only 9 inches between 150 and 300 ms. Once again, the ball drops more in the last 150 ms than in the earlier epoch, but it still follows a parabolic trajectory. Therefore, this is a curve not a break. A breaking curve ball would have to drop the same amount as this ball in the early 150-ms epoch but more than this in the last 150 ms. It would be impossible to hit the ball if it really did jump or break just before the plate. For example, a foul tip changes the ball's trajectory abruptly in the last few feet before the catcher and the umpire. As a result, these masked men cannot predict the ball's trajectory and they must wear protective clothing.

Table 14	Curve Ball Trajectories			
	80 mph		**75 mph**	
Time Since Release (ms)	Distance (ft)	Height (ft)	Distance (ft)	Height (ft)
0	55.5	6.00	55.5	6.00
50	49.7	6.20	50.0	6.18
100	44.0	6.28	44.7	6.25
150	38.3	6.24	39.4	6.20
200	32.7	6.09	34.1	6.05
250	27.2	5.83	29.0	5.78
300	21.9	5.47	23.9	5.41
350	16.5	4.99	18.8	4.93
400	11.3	4.41	13.9	4.35
450	6.11	3.73	9.0	3.67
480	3.0	**3.27**		
500			4.2	2.89
513			3.0	**2.68**

We are now ready to explain the apparent abrupt break of some curve balls. Suppose the pitcher threw the 75-mph curve ball of Table 14; it would drop 25 inches in the last 150 ms before contact. However, if the batter overestimated the pitch speed and thought it was going 80 mph, then he would expect it to fall 21 inches in the last 150 ms before contact. Thus, if he took his eye off the ball 150 ms before the projected time of contact, and saw it again when it arrived at his bat, he would think that it broke down 4 inches. Therefore, we suggest that the apparent, abrupt break of some curve balls might be a result of the batter's *overestimation* of the speed of the pitch in his mental model.

Of course, the illusion would be greatly enhanced if the batter made a mistake in estimating both the speed and the type of the pitch. If the pitch were a 95-mph fastball with 1600 rpm of backspin and the batter thought it was a 75-mph curve ball with 1900 rpm of topspin, then the illusionary jump would be 1 foot. Similarly, if such a curve ball were mistaken for such a fastball, then the illusionary break would be 1 foot.

Summary

In summary, we think every pitch (except a nonspinning pitch such as the knuckleball) follows a smooth trajectory. The apparent, abrupt jumps and drops right before the plate are perceptual illusions, caused by the batter using the optimal learning strategy of tracking and making mistakes in his mental model of the pitch. We think that in the first third of the ball's flight, the batter gathers data and forms his mental model of the pitch. In the middle third, he computes when and where the ball will cross the plate. Then he starts his swing. In the last third of the ball's flight, the batter either observes errors in his mental model so that he can track the next pitch better (the optimal learning strategy) or updates his mental model and perhaps checks his swing (the optimal hitting strategy).

One by one, scientists and engineers, using principles of physics, have explained most of baseball's peculiar pitches, for example, the knuckleball, the scuff ball and the curve ball. The most mysterious remaining pitches were the rising fastball and the breaking curve ball. Now, adding principles of psychology, we can suggest that the rising fastball is a perceptual illusion. The batter tracks the ball in the beginning, he underestimates the pitch speed and therefore incorrectly predicts the height of the ball when crosses the plate, he takes his eye off the ball to look at the predicted point of bat-ball collision and when the ball comes back into view it is higher than he thought it would be.

Generalizations

Our findings should generalize to other sports. In tennis, for example, the distances are similar, 60.5 ft for baseball and 78 ft for tennis, as are the linear velocities, 100 mph for a fast pitch and 110 mph for a fast serve. There is often a change in the ball's trajectory just before the player hits it: the baseball curves and the tennis ball bounces. Tennis coaches, even more than batting instructors, teach beginners to use the strategy with the anticipatory saccade in order to see the ball hit the racket; this strategy is probably only useful as a learning tool. Therefore, we suggest that neither baseball players nor tennis players keep their eyes on the ball. The success of the good players is due to faster smooth-pursuit eye movements, a good ability to suppress the vestibulo-ocular reflex and the occasional use of an anticipatory saccade.

Sometimes our subjects used the strategy of tracking with head and eyes but also using an anticipatory saccade, and sometimes they used the strategy of tracking with head and eyes and falling behind in the last 5 ft. Len Matin of Columbia University has speculated that athletes might use the former strategy when they are learning the trajectory of a new pitch and the latter strategy when hitting home runs.

Notes

1. A. Terry Bahill and T. LaRitz, "Why Can't Batters Keep Their Eyes on the Ball?," *American Scientist*, vol. 72, 1984, pp. 249–253. Original source of data showing how close to the plate a batter can follow the ball.

2. Ted Williams and John Underwood, *The Science of Hitting*, rev. ed., Simon and Schuster, New York, 1986. An explanation of how he did it, by one of the best hitters of all time.

3. F. Frohlich and G. R. Scott, "Where Spectators Sit to Catch Baseballs," *Baseball Research Journal*, Tenth Annual Historical and Statistical Review of the Society for American Baseball Research, vol. 10, 1981, pp. 132–138. A hard-to-find paper showing where 1000 fair and foul balls landed in 57 games in the Astrodome.

4. A. Terry Bahill and J. S. Kallman, "Predicting Final Eye Position Halfway Through a Saccade," *IEEE Transactions on Biomedical Engineering*, vol. BME-30, 1983, pp. 781–786. An engineering study of the inter- and intrasubject variability of human saccadic eye movements.

5. J. D. McDonald, A. Terry Bahill, and M. B. Friedman, "An Adaptive Control Model for Human Head and Eye Movements While Walking," *IEEE Transactions on Systems, Man, and Cybernetics*, vol. SMC-13, 1983, pp. 167–174. The first study showing eye movements while walking. It also presents a model for eye-head-body coordination while walking.

6. R. D. Lyons, "Take Your Eye Off the Ball, Scientist Coaches Sluggers," *New York Times*, June 12, 1984, pp. 19 and 23. A layman's explanation of the Bahill and LaRitz paper.

7. L. Stark, J. A. Michael, and B. L. Zuber, "Saccadic Suppression: A Product of the Saccadic Anticipatory Signal," in Christopher R. Evans and T. B. Mulholland, eds., *Attention in Neurophysiology*, Butterworth, London, 1969, pp. 281–303. The term "saccadic suppression" first appeared in this paper. It explains why we cannot see our eyes move when we look between the images of our eyes in a mirror.

8. E. Matin, "Saccadic Suppression and the Dual Mechanism Theory of Direction Constancy," *Vision Res.*, vol. 22, 1982, pp. 335–336. A psychophysical explanation of why we cannot see our eyes move when we look between the images of our eyes in a mirror.

9. Joe Brinkman, and Charlie Euchner, *The Umpire's Handbook*, Steven Greene Press, Lexington, Mass., 1986. Helpful advice for umpires.

10. D. M. Regan, "Visual Judgments and Misjudgments in Cricket, and the Art of Flight," *Perception*, vol. 21, 1992, pp. 91–115. Application of science to the game of cricket.

11. A. Terry Bahill and W. J. Karnavas, "The Perceptual Illusion of Baseball's Rising Fastball and Breaking Curveball," *Journal of Experimental Psychology, Human Perception and Performance*, vol. 19, no. 1, 1993, pp. 3–14. An explanation for the rising fastball based on psychological and physiological principles.

12. Michael McBeath, "The Rising Fastball: Baseball's Impossible Pitch," *Perception*, vol. 90, 1990, pp. 545–552. A qualitative discussion of the rising fastball with lots of references.

13. D. M. Regan, "The Eye in Ballgames: Hitting and Catching." *Sport en Zien*, 1986, pp. 7–32, Haarlem, De Vrieseborcht. How the human visual system limits sporting performance.

14. D. M. Regan, K. I. Beverley and M. Cynader, "The Visual Perception of Motion in Depth," *Scientific American*, vol. 241, no. 1, 1979, pp. 136–151. A simple explanation of visual perception of motion in depth.

15. J. T. Todd, "Visual Information About Moving Objects," *Journal of Experimental Psychology: Human Perception and Performance*, vol. 7, 1981, pp. 795–810. Discussion of the judging time to contact in different sports.

16. Fred Hoyle, *The Black Cloud*. New York, Harper, 1957. A science fiction book.

17. W. J. Karnavas, P. Sanchez and A. Terry Bahill, "Sensitivity Analysis of Continuous and Discrete Systems in the Time and Frequency Domains," *IEEE Transactions on Systems, Man, and Cybernetics*, vol. SMC-23, 1993, pp. 408–501. A complete explanation of sensitivity analyses.

Physiological Predictors of Batting Performance[1]

Many physiological parameters could be used to predict whether a particular ballplayer will be successful. To start with, it is helpful if the player is big, strong and fast. Professional teams already measure such parameters. But there are many other quantitative parameters that they could use.

Visual acuity is important for a batter. Ted Williams had 20/10 visual acuity; such acuity is possessed by about one in a thousand people in the 20- to 30-year age bracket. However, there are now more sophisticated measures of visual acuity; for example, contrast sensitivity and dynamic visual acuity could be used to assess potential. To measure acuity a person looks at letters on a chart and tries to identify them. In a test of contrast sensitivity the subject is presented a series of sine wave gratings, which look like the alternating black and white bars shown in Figure 69, and is examined to determine the closest spacing of bars that the subject can perceive. In a test of dynamic visual acuity, a letter is

Figure 69   A sine wave grating like those used to test contrast sensitivity. The bars on the left are easier to discriminate because they are wider and are more widely spaced than the bars on the right.

projected on a screen 10 feet in front of the subject. The letter is rotated clockwise starting at 100 rpm. The rotational speed is gradually decreased, and the fastest speed at which the subject can identify the letter is recorded. (This is much like trying to read the label on a rotating phonograph record. Most people can read the label at 33 rpm; few people can read it at 78 rpm.) Data collected from more than 800 subjects showed that Olympic-caliber baseball players had better static visual acuity and contrast sensitivity than age-matched nonathletes.[2,3] Professional baseball batters have superior dynamic visual acuity.[4] Tests of contrast sensitivity and dynamic visual acuity are more complicated and more time consuming than tests of static visual acuity, but they yield additional information that may be of predictive value.

Optometric exams given to the members of the 1984 U.S. Olympic volleyball team have shown big intrasubject differences in depth perception, in ability to track moving objects and in ability to not suppress vision in one eye or the other.[5,6] Getz says these differences correlate with performance.

Another physiological trait that may be correlated with good performance has to do with preference of hand and eye usage. Most humans are right-handed, although some are left-handed and a few are ambidextrous. Coincidentally most humans are also right-eye dominant. To determine your eye dominance, point to an object across the room; first close one eye and then the other. Whichever eye keeps your finger on the object is your dominant eye. (If the results of this experiment are equivocal, then instead of pointing with only one hand, form a small triangular hole between your hands, then raise both arms straight out in front of you. Sight an object across the room as before, and close first one eye then the other. Whichever eye allows you to see the object is the dominant eye.) Your finger pointing at the object across the room can only be aligned with one of your eyes. You point with your dominant eye: for most people that is their right eye. However, right-handedness does not imply right-eye dominance. Although most people are right-handed and most people are also right-eye dominant, there is no cause-and-effect link. Now comes the crux of our discussion: in the field of baseball, several investigators[7] have demonstrated that batters who are right-handed and left-eye dominant or left-handed and right-eye dominant (called cross-dominant) have statistically higher batting averages than noncross-dominant players. Although this result is statistically significant, there are no explanations for it: it is a correlation and not a cause and effect. For example, some reasons suggested for the advantage of cross-dominant batters are that their dominant eye is 2 inches closer to the pitcher, and that it is the batter's dominant "eye that is facing the pitcher"; these explanations are ridiculous. The best we can say is that cross-dominant batters have statistically higher batting averages. The cause is probably wrapped up in the complexity of the human brain and the different capabilities of each half of the brain.

Bat speed and consistency of swings are important for batting success. These parameters are already considered by the scouts. The equipment and experiments mentioned in Chapter 5 would help quantify these observations.

Quickness is another important parameter in determining potential for success. Quickness is easy to identify but hard to define. Coaches easily identify their quick players. But when asked to explain why they call these players quick, they waffle. Their uneasy verbalizations include descriptions like they react quickly, they move fast, they steal a lot of bases, they get into position to field the ball quickly, they swing the bat fast and they beat out bunts. However, all of these phrases describe results not physiological characteristics. However, we think there are physiological parameters that can be measured to determine quickness. In particular, we believe that the force-velocity curves of Chapter 5, the latency of the saccadic eye movement system and eye-hand reaction time can be used to quantify quickness. Quick players had large slopes on the force-velocity curves presented in Chapter 5. Saccadic eye movement latency can be measured with standard tests in any oculomotor laboratory. Eye-hand reaction time can be measured in the following way:

1. Holding a meter stick in front of a subject. Instructing him or her to straddle the 50-cm mark with the index finger and thumb and to watch the fingers of the experimenter, who is holding the end of the meter stick.

2. When the experimenter opens his fingers, and the meter stick begins to fall, the subject should grasp the stick between the finger and thumb.

3. The place on the meter stick where the subject catches it is an indication of eye-hand reaction time.

Thus, quickness can be measured and we think it will correlate with success playing baseball.

Many laboratories are studying the movement of athletes by digitizing television camera images with computers.[8,9] Commercially available systems cost $5,000 to $50,000. These systems record the data from a $\frac{1}{4}$ to 2-second movement in a computer. With digital reconstruction they can display a three-dimensional representation of the movement with millisecond resolution. These movements can be replayed at any speed or viewing angle, and the trajectories of individual limbs and joints can be compared to the trajectories of an expert template. Sometimes this expert template is that of a widely recognized world champion, and sometimes it is derived from the movements of the same athlete when he or she was in top form. Such studies can help an athlete change his or her motion to conform to that of the expert template and thus increase performance. Because these studies are so time consuming and expensive, they have not yet been used to predict potential, but rather they have mostly been used to help correct the movements of already highly skilled athletes.

Our final example of physiological parameters that could be measured to predict batting performance, which we will pursue in great detail, is the ability to learn to track novel visual targets. Subjects who have this ability to learn tracking of novel waveforms are more likely to be better hitters.

Good Batters Quickly Learn to Track Novel Visual Targets

Human neuromuscular systems have a time delay of 100 to 200 milliseconds (ms). To show the effects of such a time delay, hold a crisp dollar bill with George Washington's portrait between someone's outstretched finger and thumb. Tell them they can have the bill if they catch it. Then drop the bill and let them try to pinch it. Unless they make a very lucky guess, they will miss the bill, and it will drop to the floor. The human eye-tracking system has a similar time delay. The time delay in the smooth-pursuit eye-movement system is 150 ms, but humans can learn to overcome this time delay and track smoothly moving targets with no delay, provided the target position waveform is smooth, is predictable, has frequencies between 0.1 and 1.0 Hz and has small accelerations. However, it takes time to learn this zero-latency tracking. In this section, we show the time course for learning to track a particular target waveform. And we show that professional baseball players learn to track faster than typical college students do.

We measured eye movements with the photoelectric system discussed in the last chapter. However, in this study we eliminated vergence eye movements by putting the targets on a curved screen in front of the subject, and we eliminated vestibulo-ocular eye movements by immobilizing the subject's head with a head rest and a bite bar. Also, the bandwidths for the data of this chapter were larger, with the position and velocity bandwidths being 80 and 9 Hz, respectively. The subjects were two dozen male and female college students and three members of the Pittsburgh Pirates.

The visual target was a little red laser dot of light projected on a white screen in front of the subject. This dot moved leftward then rightward and the subject tried to track it, in much the same way the subject would do so seated on the net line while trying to follow the ball in a tennis match. This dot of light moved back and forth with a cubical target position waveform (shown in Figure 70).[10,11] We used this waveform because no naturally occurring visual targets move with a cubical waveform. Thus, our results were not influenced by prior learning. However, the position trace of the cubical waveform looks like the familiar sinusoid (most natural movements are sinusoidal; for example, if you jump on the bumper of a car, it will bounce up and down in a sinusoidal manner), but the velocity is strikingly different. Therefore, by analyzing the eye velocity records we could tell if the subject had really learned the cubical waveform, or if he had merely approximated it with a sinusoid. The cubical target waveform is described by the equation

$$r(t) = 10.39 \, A[2(t/T)^3 - 3(t/T)^2 + (t/T)] \text{ for } 0 < t < T$$

where T represents target period and A is the amplitude. Humans track well when the target has an amplitude of 5 degrees and a frequency of about 0.33 Hz (a 3-second period), and so these values were used in our experiments.

Figure 70 shows that a human can track the cubic target waveform very well. But this capability is not inherent. It must be learned. Our standard learning protocol began with a 6-second calibration target waveform (where the target jumped abruptly between two points 10 degrees apart), followed by 9 seconds of the cubical target waveform, 3 seconds of the square wave target waveform, another 9 seconds of the cubical waveform and finally another 6 seconds of the square wave calibration target waveform. The subjects were allowed to rest for 5 minutes and then the sequence was repeated. This process continued for about 2 hours.

Because the purpose of the eye movement system is to keep the fovea on the target, we felt that the error between the eye and the target was the most appropriate measure of the quality of tracking. Our primary metric was the mean squared error (mse) between eye position and target position. Single and double exponential curves were fit to the mse data. The best fit was usually an exponential of the form

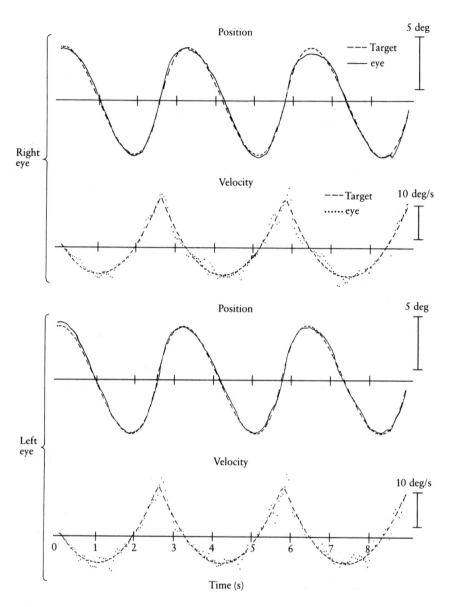

Figure 70   Position and velocity of the target and eyes as functions of time for good tracking of a cubical waveform. Upward movements in this figure represent rightward movements. (From D. E. McHugh and A. Terry Bahill, Investigative Ophthalmology and Visual Science, vol. 26, 1985, pp. 932–937.)

$$mse = Ae^{-Bt} + C$$

We called these exponential equations learning curves.

Learning Curves

Figure 71 shows learning curves for two subjects. The circles represent the human mean squared error; the solid line is the best exponential fit to the data. For the lower graph the equation for this exponential fit is

$$mse = 0.42e^{-0.26t} + 0.05$$

To give some feel for what these numbers mean, assume that the fovea (the center part of the retina that has good acuity) has a radius of 0.5 degree. Now if the target were always just on the edge of the fovea, then the mean squared error would be 0.25 deg^2. Figure 71 shows that in the beginning of the learning session the subjects were not keeping their foveae on the target, whereas at the end they were keeping the target centered on their foveae. Figure 70 shows an example of a well-practiced subject keeping the target centered on his foveae. For this figure the mean squared error between the eye and target was 0.06 deg^2.

The solid lines of Figure 72 show the exponential curves fit to the data of our four best-tracking college students. We were trying to quantify the ultimate capabilities of the human smooth-pursuit system, so we only report the performance of our best subjects. In this figure, we only show data of four of 24 college students. The other students did not demonstrate such low error tracking.

To narrow in on this exquisite tracking performance, we decided to study optimal humans performing optimally. But, who is an optimal human? For eye-tracking capability, we thought professional athletes would fit the bill. So, we invited some professional baseball players to participate in our experiments. The mse's for three members of the Pittsburgh Pirates are represented by circles, asterisks and squares in Figure 72. In viewing the target for the first time, professional baseball players 1 and 2 had much smaller mse's, 0.05 and 0.08 deg^2, than our other subjects. They had never seen a cubical waveform before, yet they started out with low mse's. Baseball players 1 and 2 each played in the major leagues for more than 10 years. Player number 3 never got out of the

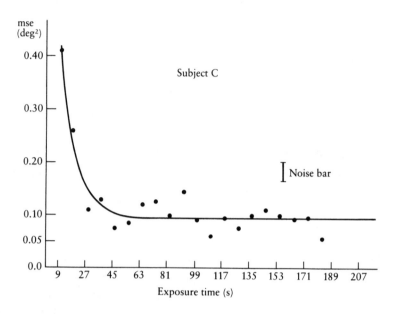

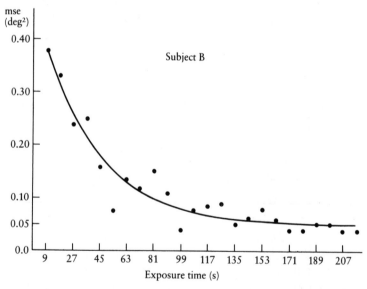

Figure 71   Time course of learning the cubical target waveform for two subjects. Mean squared error (mse) is plotted against the total time that the subject had seen the waveform. Circles are the human data points and the solid line is an exponential curve fit to the data points. (From D. E. McHugh and A. Terry Bahill, Investigative Ophthalmology and Visual Science, vol. 26, 1985, pp. 932–937.)

class-A farm system. These data seem to indicate that the ability to track the cubical waveform is correlated with baseball performance. We caution that our sample was small. However, studies with larger samples have found similar results.[12]

Why should it be important for baseball players to learn new waveforms quickly? In a typical game, a batter might see a particular pitch at most two dozen times. Given that a pitch takes about one half of a second, he would see the ball in flight for only a dozen seconds each game. Therefore, we should be studying only the first one or two data points of Figure 72, which is the region where the professional athletes excelled.

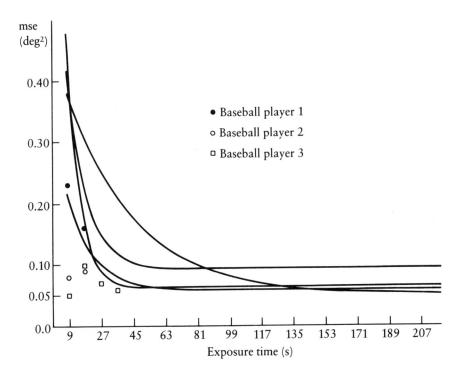

Figure 72   *Time course of learning for seven subjects. Solid lines are the exponential learning curves fit to the data of our four best students. Circles and squares are data points for three professional athletes. The professional baseball players learned the waveform faster than the students did. (Adapted from D. E. McHugh and A. Terry Bahill, Investigative Ophthalmology and Visual Science, vol. 26, 1985, pp. 932–937.)*

After a batter learns a pitch, does he retain that knowledge? The data of Figure 73 suggest that he does. Figure 73 shows the mean squared errors for one student during the initial learning session, for a session nine days later and for a session 50 days later. For each session, the data were fit with an exponential curve. Relearning was almost instantaneous. It appears that the cubical target waveform was not quickly forgotten. This suggests that the batter can also remember the trajectory of a particular pitch for a long time. Rookie pitchers and those traded from one league to another often have a period of phenomenal success. However, after they have been around a while, the batters have learned the trajectories of their pitches and the honeymoon is over.

In the laboratory, we study humans learning to track sinusoidal, parabolic and cubic waveforms. This correlates with baseball players on the field learning to track fastballs, curve balls and sliders.

Differences Between the Eyes

Humans can track certain targets very well, as shown in Figure 70. Using only smooth-pursuit eye movements, this subject was able to

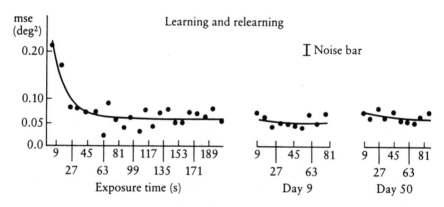

Figure 73   Time course for learning and relearning. The subject learned the waveform, returned nine days later to relearn and returned 50 days after the original experiment to relearn it again. Once people learn a certain trajectory, whether it is the trajectory of a laser beam in a laboratory or the trajectory of a pitched ball, they remember it for a long time. (From D. E. McHugh and A. Terry Bahill, Investigative Ophthalmology and Visual Science, vol. 26, 1985, pp. 932–937.)

keep his foveae on the target for more than nine seconds. Saccades were not removed or filtered out of these eye position traces; indeed small conjugate saccades can be seen at the 8.5-second mark. Please note that there are differences between the movements of the right and the left eye, as can be seen in the position traces between the 6- and 7-second marks. Comparison with the data between the 3- and 4-second marks shows that these differences are of biological origin and are not due to instrumentation errors. Differences of this magnitude between the right and left eye were typical of our data. The dominant eye did not track better than the nondominant eye; the errors were of the same magnitude. So these data show that normally there are small differences in the tracking of our two eyes; but humans are not aware of it and, therefore, it is not significant. This means that eye-tracking capabilities provide no advantage or disadvantage to being right-eye dominant and right-handed, or right-eye dominant and left-handed, or vice versa.

Generalizations

From this study, we draw the following conclusions: (1) There are differences between the tracking of the two eyes, but neither eye is consistently better; (2) professional baseball players have better tracking abilities than our best students do, specifically, they learn to track novel targets in less time.

These tests of learning to track the cubic waveform could be used in screening baseball players in the major league farm system. It could provide one more tool for evaluating potential; it could show which players have the requisite oculomotor coordination to be good hitters and which do not.

Notes

1. D. E. McHugh and A. Terry Bahill, "Learning to Track Predictable Target Waveforms Without a Time Delay," *Investigative Ophthalmology and Visual Science*, vol. 26, 1985, pp. 932–937. A study of human learning in an eye tracking task.

2. B. Coffey and A. W. Reichow, "Guidelines for Screening and Testing the Athlete's [Visual System]," *Optometric Extension Program Foundation Curriculum 11*, vol. 59, no. 7, April 1987, pp. 47–60. A protocol for optometric exams of athletes.

3. B. Coffey and A. W. Reichow, "Athletes vs. Nonathletes: Static Visual Acuity, Contrast Sensitivity, Dynamic Visual Acuity," *ARVO Abstracts,* vol. 19, 1989, p. 517. A preliminary report of their data.

4. H. Solomon, W. J. Zinn and A. Vacroux, "Dynamic Steroacuity: a Test for Hitting a Baseball," *Journal of the American Optometric Association,* vol. 59, no. 7, 1988, pp. 522–526.

5. D. J. Getz, "Vision and Sports," *Journal of the American Optometric Association,* vol. 49, 1978, pp. 385–388. An optometric study of Olympic volleyball players.

6. J. Hansen, "For Their Eyes Only," *Volleyball Monthly,* May 1983, pp. 23–25.

7. J. M. Portal and P. E. Romano, "Patterns of Eye-Hand Dominance in Baseball Players," *New England Journal of Medicine,* vol. 319, 1988, pp. 655–656. A short summary of data gathered from a college baseball team.

8. B. Cramblitt, "Computers Capture Moments of Motion," *Computer Graphics World,* March 1989, pp. 50–58. Explanation of computer-TV camera systems for studying motion of nonbaseball athletes.

9. T. S. Perry, "Biomechanical Engineered Athletes," *IEEE Spectrum,* vol. 27, no. 4, 1990, pp. 43–44. Explanation of computer systems for studying movements of athletes.

10. A. Terry Bahill and J. D. McDonald, "Model Emulates Human Smooth Pursuit System Producing Zero-Latency Target Tracking," *Biological Cybernetics,* vol. 48, 1983, pp. 213–222. The mental model used for visual tracking of moving objects.

11. A. Terry Bahill and J. D. McDonald, "Smooth Pursuit Eye Movements in Response to Predictable Target Motions," *Vision Research,* vol. 23, 1983, pp. 1573–1583. Eye movements of humans tracking moving targets. This contains the data used to make the mental model for eye tracking.

12. K. M. Daum, and D. A. Corliss, "Eye Movements of Athletes," *Investigative Ophthalmology and Visual Science* (supp), vol. 31, 1990, p. 93.

Baseball Batting Statistics:

Were Yesterday's Players Better?

very baseball fan knows that nobody has batted .400 since Ted Williams turned the trick in 1941. At any given moment only a half dozen players in the combined major leagues have batting averages over .333. Is this because the old-timers were better hitters? Is it because more hitters today are swinging for the fences? Are modern-day pitchers better?

First, let us look at the data. Figures 74, 75 and 76 show us that both the game and the players have changed over the years. The number of double plays per game has increased somewhat, and the number of errors per game has decreased markedly over the years (Figure 74). This implies that the abilities of fielders have improved. Figure 75 shows that both the number of home runs per game and the number of strikeouts per game have increased. Both began their most recent steady increase in the 1920s, and strikeouts appear to have leveled off since about 1960. Home runs per game have recently begun another increase. Because these changes are in close correlation, the "swinging for the fences" theory of increased home run production gains credence. Batting averages and

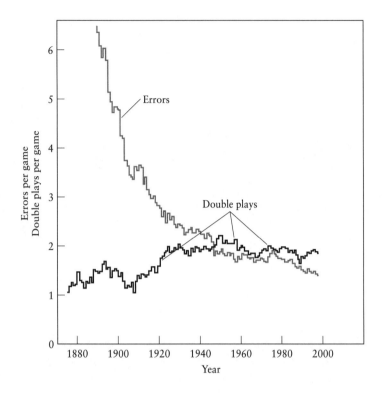

Figure 74 Figure 74   *Double plays and errors per game.*

earned runs per game, on the other hand, have remained relatively constant over the years (Figure 76). To be sure, there have been periods of increase and decrease in these figures, but, ultimately, batting averages hover around .260 and earned runs per game remain close to 3.8. In 1941, when Ted Williams batted .406, the average batting average in the combined major leagues was .262 and 3.89 earned runs were produced per game. Perhaps this means that Williams was far better than the average player of his day, while Tony Gwynn, admittedly a superstar, is not quite so far above today's average player. Baseball fans, however, still like to ask the ultimate question: Were Ty Cobb, Babe Ruth and Ted Williams better hitters than, say, Tony Gwynn, Mike Piazza and Wade Boggs?

John Thorn and Pete Palmer[1] have described a technique for making such comparisons, concluding that many of the best modern-day players would have been superstars of even greater proportions had

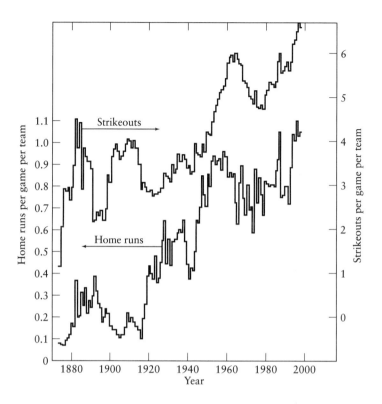

Figure 75   Home runs and strikeouts per game, per team.

they played the game 50 years before with the same abilities they possess today. Thorn and Palmer pointed to a study by R. Cramer,[2] where a method was suggested for comparing the batting skills of players in different years. Cramer calculated the Batter Win Average, a measure of runs produced above the average, divided by the number of plate appearances by the player. He then converted these results to predicted batting averages. Cramer's method leads to some astonishing predictions, as pointed out by W. Rubenstein.[3] For example, the "real lifetime batting averages" of yesteryear's stars are very low. While George Brett and Rod Carew had "real lifetime batting averages" of .341 and .325 (the best of the group), Ty Cobb batted only .289 and Babe Ruth, a paltry .262. Rubinstein predicted that "the problem with these statistics is that no one will believe them." Should they? Rubinstein further stated:

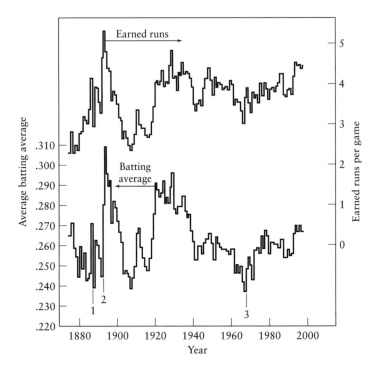

Figure 76   Batting averages and earned runs per game, per team.

Dallas Adams has pointed out that the table of "corrections" Cramer has compiled applies to an entire league in a particular season, not to a particular player. And while one might well have to subtract 101 points from the batting average of a mediocre American League hitter in 1921 to arrive at his 1976 batting average, is it really true of the Ty Cobbs and George Sislers? Evidence that it is not so seems implied in the well documented fact that the gap between the average league batting average and the league-leading batting average has been consistently narrowing throughout the century: in 1911, for instance, Cobb's batting average exceeded the American League average by 137 points, but most recent batting champs have exceeded their league batting average by only 60 to 70 points. It thus seems that the very best hitters of the past were very much better than the average hitters of their day. Because of this, Cramer's "corrections" probably cannot be applied to all players of the past: they almost certainly cannot be applied to the very best players of former ages.

This is a jewel of insight. "Average" hitters of today may be very much better than "average" old-timers, but perhaps the very best players of both eras are comparable in skill.

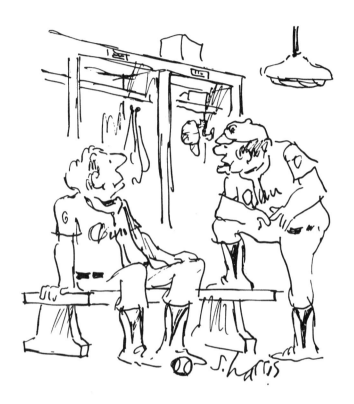

"I'M SORRY, ROGER, BUT YOU'VE REACHED THE MANDATORY RETIREMENT BATTING AVERAGE OF .150."

S. J. Gould[4] has proposed a remarkably simple and, at first glance, counterintuitive explanation of the demise of the .400 hitter by proceeding precisely along the line of reasoning first expounded by Rubinstein. Gould claims that while the best players of today are neither appreciably better nor worse than the best of the old-timers, there are more very good players now. But wouldn't this mean more, rather than fewer, .400 hitters? The theory requires some explanation.

Records are made to be broken. There is little doubt that athletes get better over the years. People in general are healthier and bigger than they were 50 years ago. There are also more people actively competing in various sports than there were in the not too distant past. To see that the top athletes get better and better one only needs to look at the record book for Olympic sports like running, jumping, throwing and swimming. Today, the sub-four-minute mile is commonplace. Before the 1950s, it was considered a virtual impossibility. As a specific example of how athletic improvement has occurred over the years, consider the Boston Marathon. In 1922, Clarence DeMar ran the Boston Marathon in 2:18:10, a record that stood for 34 years, when Aniti Viskari of Finland ran the 26-mile distance in 2:14:14. Today, the record stands at 2:07:15 (set by Cosmas Noleti of Kenya in 1994), a mere 7.5 percent improvement over DeMar's time. In the years between 1921 and 1956, there were some remarkably slow race winners (Figure 77). In 1931, the winning time was 2:46:45.8. Since 1969, no winning time has been above

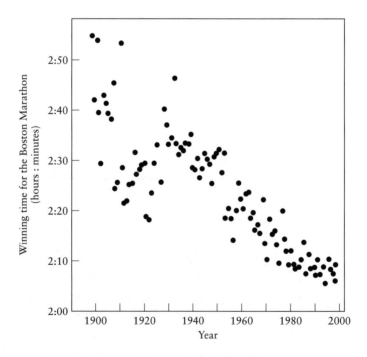

Figure 77   *Running times for the Boston Marathon.*

2:19. Only runners who have run a marathon in less than 3:30 during the previous year are allowed to participate in the race. Thousands qualify. Scores of runners finish in less than 2:25 each year. The conclusion seems clear. The very best runners today, in spite of vast improvements in training techniques, diet, equipment and so on, are better, but not tremendously better, than the best runners of 50 or 60 years ago. As any serious road racer will testify, however, there are many more exceptionally good runners.

Baseball is a very different sport from running. In running, the athlete is pitted against the clock. To be sure, other people are involved and the aim is to defeat them, but one is not allowed to interfere with their efforts. The faster and stronger a runner is, the more likely he or she is to win and to set a new record. In baseball, things are different. The batter's job is to hit the ball and to reach base safely. The pitcher's job (along with that of his teammates) is to make the batter's job as difficult as possible. If it makes sense that the skill of the average batter should have improved over the years, then it makes as much sense that the skill of the average pitcher has also improved. Which has improved more? (As a former pitcher, one of us [Watts] must admit that it seems that pitching appears to have a historic tendency to dominate the game, but this is not important.) What is important is that whenever one side, offense or defense, begins to hold the upper hand and to rule the game, the spectators begin to lose interest and the owners change the rules to restore equilibrium.

Baseball is a finely tuned sport, and only the most delicate adjustments are required to bring the game into balance when it strays. In fact, one can easily recognize the effects of rule changes in Figures 75 and 76. Some rule changes that produced detectable changes are listed in Table 15. The simple bookkeeping change of not counting a base on balls as a hit, for example, shows up dramatically. The rule was instituted in 1888, and the average batting average dropped from .271 the previous year to .239 (see the first arrow in Figure 76). The following year the number of balls necessary to walk a batter was decreased from seven to four. The number of walks per nine innings increased from 2.21 in 1888 to 3.46 in 1889. The pitcher also had to "groove" the ball in order to avoid walking even more batters, and batting averages responded by increasing to .263.

In 1893, the pitching distance was increased from 50 feet to 60.5 feet. Batting averages increased from .245 in 1892 to .290 in 1893 (second arrow in Figure 76) and to .309 in 1894. They then began a slow downward drift. Spitball pitching began, and perhaps knuckleballs and other trick

Table *15*	Some Rule Changes over the Years
1888	Strike out on 3 strikes instead of 4.
	Base on balls not counted as a hit.
1889	Base on balls on 4 balls instead of 6.
1893	Pitching distance moved from 50 ft to 60.5 ft.
1894	Batter charged with strike for foul bunt.
1895	Bat diameter increased from 2.5 to 2.75 in.
	Strike charged for foul tip.
1903	Pitcher's mound no higher than 15 in.
1909	Foul bunt on third strike is a strikeout.
	Cork centered ball introduced.
1920	Spitball abolished (actually gradually).
1926	Sacrifice fly introduced.
	Ground rule double.
1940	No sacrifice fly.
1954	Sacrifice fly reintroduced.
1963	Strike zone enlarged.
1969	Strike zone shrunk.
	Pitcher's mound lowered to 10 in.

pitches were introduced. Pitchers slowly began to dominate. By 1908, the average batting average had dropped to .239 and the number of earned runs per game per team had dipped from 5.32 in 1894 to 2.37. In an effort to stem the tide, the cork-centered ball was introduced near the end of the 1909 season. The hitters responded by batting .250 and scoring an average of 2.78 earned runs per game per team in 1910, and these numbers climbed to .269 and 3.37 by 1912. But by 1918, they had again dropped. In that year, the batting average was .254 and only 2.77 earned runs were scored per each team game. Something else had to be done.

In 1919, the use of the spitball and other trick pitches like the shine ball and the emery ball were prohibited. Actually, the rule did not take effect until the 1920 season, and even then a "grandfather clause" was adopted, first allowing each team to name two spitball pitchers for the 1920 season and then, in 1921, allowing eight National League and nine American League pitchers to use the pitch for the remainder of their careers. Hitting dominated for the next 10 to 20 years, with batting averages hovering around .285 and earned runs per game per team between

4.0 and 5.0. In 1930, the average batting average was .296 and teams earned an average of 4.81 runs per game. Home runs per game skyrocketed from 0.23 in 1918 to 1.26 (or 0.63 per team) in 1930.

By 1940, the pitchers were beginning to come into their own again. In 1955, hitters batted .259 and teams produced only 4.00 earned runs per game. By 1968, these numbers had fallen to .237 and 2.98. The owners began to get nervous. The 1969 season began with two major rule changes in place. The strike zone was to include only the area between the armpit and the top of the knees (it has previously been from the top of the shoulders to the bottom of the knees), and the height of the pitcher's mound was lowered from 15 inches to 10 inches.

The response was immediate. The average major league batting average rose to .248 in 1969 (third arrow in Figure 76) and to .254 in 1970. Earned runs per game per team rose to 3.62 and then to 3.89. The numbers jumped again slightly in 1977, when the American League introduced the designated hitter, but have otherwise remained fairly constant.

If seemingly minor rule changes are reflected in average performances, how do these show up in the performances of individuals? Every individual is different, of course. Ted Williams, Ty Cobb, Rod Carew, Wade Boggs and Tony Gwynn were all different people with different personalities who played the game in different home parks under different conditions and against different opponents. We will argue, as did Gould, that a reasonable measure of an individual hitter's performance in a given year is the number of standard deviations above the league average the hitter batted during that year.

The argument, elaborated by Gould, is as follows: Suppose the abilities of the very best players are invariant, or nearly invariant, from player to player over different eras. These are the Cobbs, Bretts and Gwynns of any era. The rest of the players making up a league have abilities spread over a spectrum. Suppose we are to select 500 players to make up a baseball league. The larger the pool from which the players are drawn, the narrower will be the spectrum of abilities. Imagine, for example, a society so large that it could produce an entire league of players who could bat as well as Ty Cobb. If every player had an average year, the league batting average would be, say, .360, and the range of batting averages would be zero. Everyone would hit .360. If the rules were now changed to favor the pitcher sufficiently to bring the league average down to .260 (all pitchers like Walter Johnson), there would, of course, be no .300 hitters, only an entire league full of .260 hitters.

In the real world, of course, the abilities of hitters in the major leagues are still spread over a spectrum. The talent pool has, however, increased over the years for a variety of reasons. In addition to population growth itself, the entry of blacks, Mexicans, Puerto Ricans and others has considerably increased the potential number of outstanding players. One might therefore expect a decrease over time in the range of batting averages in a given year in the major leagues. Such a decrease has, in fact, occurred.

We have calculated the means and the standard deviations for each of the major leagues from each league's inception through 1998. The results are shown as Tables 16 and 17. We used data tabulated by Joe Reichler and from the World Wide Web site www/.baseball stats.com. In performing the calculations, only players having at least two at-bats per game on the average were included, so that the results would reflect the abilities of "regular" players. Gould has provided us with the results of a similar calculation that he performed, including players with at least one at-bat per game. Gould's results are in parentheses in Tables 16 and 17. His calculated standard deviations are somewhat larger than ours, reflecting the fact that players with between one and two at-bats per game were of marginal ability. The trends in the two sets of results are quite similar. Along with considerable year-to-year scatter, one does see the expected gradual decrease in standard deviation with time.

Table 16 National League Batting Averages

Year	Mean	Standard Deviation	Gould's Standard Deviation
1876	.26547	.04976	
1877	.27540	.04783	
1878	.25934	.05210	
1879	.25612	.04837	
1880	.24855	.03998	
1881	.26264	.04386	
1882	.25813	.04097	

Table 16 National League Batting Averages (continued)

Year	Mean	Standard Deviation	Gould's Standard Deviation
1883	.26485	.04163	
1884*	.25294	.04353	
1885	.25343	.04693	
1886	.26340	.04794	
1887	.27808	.03963	
1888	.25245	.03665	
1889	.27649	.03678	
1890**	.26418	.03547	
1891	.26263	.02919	
1892	.25809	.03301	
1893	.29065	.03647	
1894	.32098	.04147	
1895	.30764	.04302	
1896	.30424	.03971	
1897	.30188	.04178	
1898	.28188	.03740	
1899	.29949	.03793	
1900	.29353	.03593	
1901	.28590	.04491	(.0474)
1902	.27056	.03920	(.0443)
1903	.28537	.03535	(.0393)
1904	.26343	.03121	(.0375)
1905	.27280	.03476	(.0382)
1906	.25884	.03575	(.0368)
1907	.25763	.02933	(.0305)
1908	.25308	.03120	(.0342)
1909	.25726	.03141	(.0349)
1910	.26829	.02854	(.0328)
1911	.27367	.03255	(.0361)
1912	.28824	.02747	(.0361)
1913	.27667	.02597	(.0281)
1914	.26731	.02852	(.0310)
1915	.26186	.02594	(.0292)
1916	.26189	.02820	(.0318)
1917	.26458	.02638	(.0298)

Table 16 National League Batting Averages (continued)

Year	Mean	Standard Deviation	Gould's Standard Deviation
1918	.26913	.02889	(.0365)
1919	.27737	.02390	(.0359)
1920	.28150	.03039	(.0339)
1921	.30495	.02993	(.0344)
1922	.30548	.03119	(.0381)
1923	.29928	.03265	(.0349)
1924	.29492	.03375	(.0352)
1925	.30805	.03167	(.0333)
1926	.29593	.02831	(.0367)
1927	.29569	.02974	(.0338)
1928	.29807	.03448	(.0352)
1929	.31062	.03515	(.0397)
1930	.31715	.03591	(.0389)
1931	.29176	.03265	(.0357)
1932	.29137	.03190	(.0359)
1933	.28134	.03106	(.0337)
1934	.29267	.02969	(.0308)
1935	.29112	.03061	(.0290)
1936	.29279	.03350	(.0341)
1937	.28355	.03912	(.0368)
1938	.27872	.02802	(.0289)
1939	.28605	.02717	(.0312)
1940	.27939	.02879	(.0328)
1941	.27230	.02904	(.0329)
1942	.26202	.02621	(.0304)
1943	.26948	.03233	(.0362)
1944	.27637	.03187	(.0343)
1945	.27969	.02719	(.0312)
1946	.27113	.02869	(.0303)
1947	.28115	.02512	(.0328)
1948	.27566	.02968	(.0341)
1949	.28236	.02680	(.0317)
1950	.28053	.02765	(.0306)
1951	.27748	.02820	(.0334)
1952	.26703	.02876	(.0324)

Table 16 National League Batting Averages (continued)

Year	Mean	Standard Deviation	Gould's Standard Deviation
1953	.28217	.03240	(.0347)
1954	.28241	.03324	(.0360)
1955	.27496	.02868	(.0319)
1956	.27063	.02858	(.0315)
1957	.27498	.02829	(.0339)
1958	.27875	.02871	(.0314)
1959	.27828	.02689	(.0311)
1960	.27180	.02453	(.0308)
1961	.27386	.03038	
1962	.27592	.02787	(.0295)
1963	.26141	.03164	(.0324)
1964	.27057	.03093	(.0326)
1965	.26630	.02815	(.0334)
1966	.27388	.02603	(.0316)
1967	.26717	.03360	(.0370)
1968	.25870	.02916	(.0311)
1969	.26607	.03170	(.0350)
1970	.27517	.03380	(.0342)
1971	.26779	.03340	(.0344)
1972	.26481	.03197	(.0348)
1973	.27088	.02895	(.0344)
1974	.27117	.02652	(.0303)
1975	.27379	.03204	(.0348)
1976	.27275	.03088	(.0377)
1977	.27611	.02557	(.0312)
1978	.26945	.02489	(.0305)
1979	.27352	.02638	(.0303)
1980	.27021	.02540	(.0315)
1981	.26991	.03132	(.0315)
1982	.27089	.02342	(.0361)
1983	.26934	.02531	
1984	.26864	.02614	
1985	.26840	.02525	
1986	.26623	.02815	
1987	.27853	.02616	
1988	.26021	.02415	

Table 16 National League Batting Averages (continued)

Year	Mean	Standard Deviation	Gould's Standard Deviation
1989	.26228	.02930	
1990	.27193	.02758	
1991	.25141	.02239	
1992	.26307	.03000	
1993	.27778	.02956	
1994	.27603	.03355	
1995	.27687	.03167	
1996	.27635	.02978	
1997	.27244	.02949	
1998	.27696	.03080	
1999			

*Union Association: .26116; .05019.
**Player's League: .28449; .04321.

Table 17 American League Batting Averages

Year	Mean	Standard Deviation	Gould's Standard Deviation
1882	.24880	.04717	
1883	.25869	.03629	
1884	.25490	.04130	
1885	.25713	.04299	
1886	.25663	.03773	
1887	.28402	.04430	
1888	.24788	.03658	
1889	.27110	.03784	
1890	.26880	.04648	
1891	.26601	.03845	
1901	.29003	.03355	(.0389)
1902	.29110	.03662	(.0421)

Table 17	American League Batting Averages (continued)		
Year	Mean	Standard Deviation	Gould's Standard Deviation
1903	.26905	.03612	(.0399)
1904	.25593	.03459	(.0360)
1905	.25183	.03114	(.0363)
1906	.26395	.03680	(.0438)
1907	.26142	.02917	(.0381)
1908	.25045	.02856	(.0328)
1909	.26053	.03693	(.0372)
1910	.26175	.03371	(.0424)
1911	.29293	.04351	(.0483)
1912	.27998	.04100	(.0424)
1913	.27708	.03860	(.0411)
1914	.26495	.03519	(.0382)
1915	.26446	.03576	(.0371)
1916	.26281	.03633	(.0354)
1917	.26087	.03946	(.0398)
1918	.27269	.03803	(.0391)
1919	.28260	.03539	(.0398)
1920	.29766	.03906	(.0431)
1921	.30541	.03208	(.0430)
1922	.29990	.03583	(.0373)
1923	.29518	.03875	(.0431)
1924	.30547	.03110	(.0354)
1925	.30610	.03897	(.0435)
1926	.29658	.03495	(.0366)
1927	.30097	.03679	(.0366)
1928	.29702	.03286	(.0369)
1929	.29872	.03429	(.0363)
1930	.30917	.03490	(.0393)
1931	.29564	.03311	(.0346)
1932	.29182	.03048	(.0344)
1933	.28592	.02994	(.0336)
1934	.29878	.02968	(.0371)
1935	.29319	.02516	(.0321)
1936	.30340	.03307	(.0357)
1937	.29707	.03252	(.0371)
1938	.29502	.02890	(.0335)

Table *17*   American League Batting Averages (continued)

Year	Mean	Standard Deviation	Gould's Standard Deviation
1939	.29542	.02800	(.0318)
1940	.28434	.03298	(.0355)
1941	.28194	.03401	(.0365)
1942	.27466	.02727	(.0321)
1943	.26067	.02593	(.0303)
1944	.27428	.02721	(.0314)
1945	.27035	.02413	(.0298)
1946	.27420	.03338	(.0346)
1947	.27059	.03052	(.0349)
1948	.27954	.03195	(.0320)
1949	.27835	.02599	(.0298)
1950	.28437	.03193	(.0314)
1951	.27613	.02583	(.0340)
1952	.26860	.02630	(.0270)
1953	.27476	.02569	(.0302)
1954	.27012	.03084	(.0319)
1955	.27398	.02893	(.0365)
1956	.27453	.03272	(.0346)
1957	.27012	.03441	(.0344)
1958	.27409	.02857	(.0332)
1959	.26756	.03058	(.0309)
1960	.27268	.02149	(.0268)
1961	.27349	.02665	(.0293)
1962	.26855	.02629	(.0300)
1963	.26022	.02381	(.0269)
1964	.26330	.02360	(.0272)
1965	.25635	.02553	(.0285)
1966	.25457	.02246	(.0276)
1967	.25043	.02760	(.0306)
1968	.24827	.02634	(.0333)
1969	.26596	.02467	(.0305)
1970	.26301	.02983	(.0318)
1971	.26579	.02534	(.0306)
1972	.25593	.02606	(.0306)
1973	.26709	.02650	(.0295)
1974	.26450	.02764	(.0283)

Table 17	American League Batting Averages (continued)		
Year	Mean	Standard Deviation	Gould's Standard Deviation
1975	.26624	.02818	(.0325)
1976	.26348	.02850	(.0310)
1977	.27243	.03115	(.0325)
1978	.26980	.02227	(.0293)
1979	.27476	.02834	(.0289)
1980	.27772	.02965	(.0324)
1981	.26116	.03016	(.0326)
1982	.27194	.02625	
1983	.27101	.02956	
1984	.26979	.03012	
1985	.26763	.02636	
1986	.26852	.02838	
1987	.27181	.03005	
1988	.26359	.03092	
1989	.26482	.02854	
1990	.26261	.02724	
1991	.26353	.03246	
1992	.25920	.03907	
1993	.27088	.03149	
1994	.27772	.03288	
1995	.27484	.03057	
1996	.27940	.03144	
1997	.27423	.02913	
1998	.27628	.02824	
1999			

When we do statistical analyses, the larger the group of subjects, or tests, the more meaningful the analysis becomes. Generally, when the number of subjects is very large, the data begins to look like a standard distribution and the standard deviation takes on a specific meaning. In simple terms, about 68 percent of the subjects fall within the range of one standard deviation about the average, 95 percent fall within two standard deviations and 99.7 percent fall within the range of three standard deviations. A batter might be considered to have an outstanding year if his average places him more than two standard deviations above the league average. If he hit more than three standard deviations above the average, he would qualify as a genuine superstar. The number of standard deviations n above the league average that an individual player achieved can be calculated using the simple formula

$$n = \frac{\text{individual batting average } - \text{ league batting average}}{\text{standard deviation for that year}}$$

In calculating n, we used the league batting average, as tabulated in Tables 16 and 17, rather than that compiled by Reichler. This league batting average as well as the standard deviation therefore reflect "regular" players—those who batted two or more times during an average game. Table 18 is a summary of the results. This table lists all those players whose batting average in a given year was more than three standard deviations above the league average.

The first two entries will not surprise baseball aficionados. Napoleon Lajoie's n of 3.934 occurred when he hit .422 during 1901 when the American League average was .290 and the standard deviation was 0.034. In 1924, Rogers Hornsby batted 3.825 standard deviations above the National League average. The third, fourth, and fifth entries will surprise many. Because both batting averages and standard deviation have been generally low in recent years, an individual player's batting average need not reach the lofty heights of the .400 hitters of earlier days to be more than three standard deviations above the league average. Both George Brett and Rod Carew turned the trick while batting in the .380s, while Wade Boggs' .368 placed him third on the all-time best single season list. In fact, Carew appears on the list three more times, with batting averages of "only" .364, .359 and .350. Boggs appears five times in seven years and missed a sixth year only very narrowly. Tony Gwynn appears four times, narrowly missing several times.

Table 18 contains a smattering of players from all eras, which suggests that the best players of every era were about equal in ability. Note, however, that some of the great players of the past are surprisingly absent. Conspicuous for its absence is the name of "Shoeless" Joe Jackson, who compiled a lifetime batting average of .356, batting .408 in 1911 and .395 in 1912. The trouble is that during those two years, the American League batting averages were .293 and .280, and the standard deviations

Table 18   All-time Highest n for Individual Years

Player	Year	Batting Average	n
N. Lajoie	1901	.422	3.934
R. Hornsby	1924	.424	3.825
W. Boggs	1985	.368	3.808
G. Brett	1980	.390	3.787
R. Carew	1977	.388	3.710
T. Cobb	1910	.385	3.656
T. Williams	1941	.406	3.648
N. Lajoie	1910	.384	3.627
N. Lajoie	1904	.381	3.616
H. Stovey	1884	.404	3.610
R. Carew	1974	.364	3.600
T. Gwynn	1994	.394	3.516
T. Gwynn	1987	.370	3.497
T. Williams	1957	.388	3.426
T. O'Neill	1887	.435	3.408
T. Speaker	1916	.386	3.391
S. Musial	1948	.376	3.381
G. Sisler	1922	.420	3.352
W. McGee	1985	.353	3.350
W. Boggs	1988	.366	3.312
R. Carew	1975	.359	3.292
N. Cash	1961	.361	3.284
S. Musial	1946	.365	3.272
H. Walker	1947	.363	3.258
R. Barnes	1876	.429	3.256
H. Wagner	1908	.354	3.235
L. Walker	1997	.366	3.173
T. Cobb	1912	.410	3.171
T. Cobb	1909	.377	3.154
T. Gwynn	1984	.351	3.151
H. Wagner	1907	.350	3.149
R. Carew	1973	.350	3.129
A. Galarraga	1993	.370	3.120
W. Boggs	1986	.357	3.118
W. Keeler	1897	.432	3.114
C. Anson	1881	.399	3.109
T. Cobb	1917	.383	3.095

Player	Year	Batting Average	n
Table 18 All-time Highest *n* for Individual Years (continued)			
R. Garr	1974	.353	3.086
R. Hornsby	1921	.397	3.076
T. Cobb	1922	.401	3.073
A. Vaughn	1935	.385	3.067
R. Hornsby	1922	.401	3.063
J. Dimaggio	1939	.381	3.056
H. Zimmerman	1912	.372	3.049
W. Boggs	1983	.361	3.044
T. Cobb	1907	.350	3.037
W. Boggs	1987	.363	3.035
M. Piazza	1997	.362	3.034
F. Dunlap	1884	.412	3.005

were 0.044 and 0.410. Even Ty Cobb's .420 batting average in 1911 placed him less than three standard deviations (2.920) above the league average. Hugh Duffy's all-time high batting average of .438 in 1894 does not qualify for Table 18. Three other players also hit .400 or better during that year. The National League batting average was .321 and the standard deviation 0.041. Hitting three standard deviations above the league average would have required a player to bat .446!

We have not yet computed lifetime statistics of most of the players. We have done this for 10 players in Table 19: Wade Boggs, Tony Gwynn, George Brett, Rod Carew, Ty Cobb, Rogers Hornsby, "Shoeless" Joe Jackson, Napoleon Lajoie, Babe Ruth and Ted Williams. Although Brett appears fourth on the all-time best single season list, his lifetime average value of *n* is "only" 1.513. Jackson has no single season better than 3.0, but his consistency helped him to gain a 2.054 lifetime *n*. Babe Ruth was a slugger. The *n* of 1.569 does not, of course, accurately reflect his value as a hitter. As a power hitter, he was, until recently, without peer.

Carew and Hornsby show 2.125 and 2.230 over long careers, while Williams has an impressive 2.137, showing considerable consistency over a career interrupted by two wars. The most impressive numbers for those having completed long careers belong, as might be expected, to Ty Cobb.

Table 19 Career n's for 10 Players

Tony Gwynn		Ty Cobb		Rogers Hornsby		Ted Williams		Rod Carew	
Year	n	Year	n	Year	n	Year	n	Year	n
1982	0.733	1906	1.523	1916	1.812	1939	1.128	1967	1.506
1983	1.567	1907	3.037	1917	2.366	1940	1.809	1968	0.903
1984	3.151	1908	2.572	1918	0.411	1941	3.648	1969	2.677
1985	1.925	1909	3.154	1919	1.700	1942	2.983	1970	1.626
1986	2.230	1910	3.656	1920	2.912	U.S. Air Force		1972	2.382
1987	3.497	1911	2.920	1921	3.076	1946	2.031	1973	3.129
1988	2.186	1912	3.171	1922	3.063	1947	2.373	1974	3.600
1989	2.516	1913	2.925	1923	2.595	1948	2.800	1975	3.292
1990	1.344	1914	2.928	1924	3.825	1949	2.487	1976	2.369
1991	2.929	1915	2.923	1925	2.998	1950	1.022	1977	3.710
1992	1.798	1916	2.978	1926	0.744	1951	1.621	1978	2.838
1993	2.714	1917	3.095	1927	2.196	U.S. Air Force		1979	1.526
1994	3.516	1918	2.874	–	–	1954	2.428	1980	1.797
1995	2.877	1919	2.865	1931	1.202	1955	2.835	1981	1.454
1996	2.574	1920	0.930	Career	2.230	1956	2.154	1982	1.793
1997	3.376	1921	2.606			1957	3.426	1983	2.300
1998	1.428	1922	2.822			1958	1.887	1984	0.837
Career	2.374	1923	1.157			1959	−0.443	1985	0.469
		1924	1.046			Career	2.137	Career	2.125
		1925	1.845						
		1926	1.214						
		1927	1.523						
		1928	0.791						
		Career	2.372						

Joe Jackson		Wade Boggs		Nap Lajoie		Babe Ruth		George Brett	
Year	n	Year	n	Year	n	Year	n	Year	n
1911	2.645	1982	2.936	1897	1.463	1918	0.718	1974	0.633
1912	2.805	1983	3.044	1898	1.233	1919	1.113	1975	1.482
1913	2.485	1984	1.833	1899	2.123	1920	2.006	1976	2.439
1914	2.076	1985	3.808	1900	1.460	1921	2.263	1977	1.270
1915	1.218	1986	3.118	1901	3.934	1922	0.421	1978	1.087
1916	2.192	1987	3.035	1902	2.045	1923	2.524	1979	1.914
1917	1.017	1988	3.312	1903	2.380	1924	2.332	1980	3.787
1919	1.933	1989	2.284	1904	3.616	1925	−0.413	1981	1.752
1920	2.159	1990	1.446	1905	2.478	1926	2.158	1982	1.107

Table 19 Career n's for 10 Players (continued)

Joe Jackson		Wade Boggs		Nap Lajoie		Babe Ruth		George Brett	
Year	n	Year	n	Year	n	Year	n	Year	n
Career	2.054	1991	2.109	1906	2.474	1927	1.496	1983	1.319
		1992	0.000	1907	1.288	1928	0.791	1984	0.472
		1993	0.988	1910	3.627	1929	1.350	1985	2.556
		1994	1.955	1911	1.656	1930	1.428	1986	0.757
		1995	1.608	1912	2.147	1931	2.336	1987	0.605
		1996	1.005	1913	1.501	1932	1.614	Career	1.513
		1997	0.610	1914	−0.197	1933	0.504		
		1998	0.132	1915	0.435	1934	−0.363		
		Career	1.954	1916	−0.463	Career	1.569		
				Career	1.813				

For one thing, his career spanned 23 years. Between 1907 and 1922 he hit less than 2.6 standard deviations above his league average only once. For consistent batting excellence over an extended period, Cobb was apparently without peer.

But watch out for Tony Gwynn! He is hitting an astonishing 2.374 standard deviations above the National League average. If he continues at this pace, he may well replace the Georgia Peach as the greatest hitter of all time.

Homers Galore!

Even in the shadow of the unforgettable year of 1998 and the race for the home run record by Mark McGwire and Sammy Sosa, 1996 still stands out as the record year for home runs per team per game. The year 1987 also saw a surge in home runs, from 0.91 in 1986 to 1.05 in 1987. There was a succession of very warm years in the mid-1980s. This may have contributed to the surge of home runs. By the late 1980s, the weather in the East and Midwest had cooled a bit and home run production decreased and remained relatively constant. But 1993 was an expansion year for the National League and the Colorado Rockies and the Florida Marlins joined the league. Home runs started flying in Colorado's mile-high stadium with its thin air. In 1993, the Rockies set a record when opponents scored 967 runs. They led the league in runs scored against them from 1993 to 1997, in runs scored by them in 1995, 1996 and 1997, in home runs in 1995, 1996 and 1997, and in home runs given up in 1993, 1996 and 1997. The Florida Marlins stadium in Miami is also a hitter's park. In 1992, Camden Yards opened for the Baltimore Orioles. It is a very good park for hitting home runs. The seats along the sidelines are unusually close to the diamond, so that fewer pop-ups are caught, giving the batter maybe a few extra swings. All these things have contributed to the increase in home runs.

We continue to feel that there is no credible reason to believe that the ball is livelier. Many things have contributed to the increase in home runs, the most important being the livelier ballplayer. There are several matters here. To understand the first one, one must only look at the arms of some of the players. Mark McGwire's arms are the size of most men's legs! If he hits the ball solidly, no park can contain the ball. Remember that in Chapter 5 we discussed the ideal bat weight. A glance at Figure 43, for example, shows that while ideal bat weights for these individuals are around 33 ounces, the graph is flat in that range. Many players are choosing a lighter bat without losing much in the way of batted-ball speed. Choosing a lighter bat increases the batter's ability to control the bat, which allows the hitter to meet more pitches squarely. The second matter is that fans love the long ball. Many hitters look for home runs every time they come to the plate. One does not see McGwire or Sosa hitting behind the runner or laying down a sacrifice bunt.

Nor did Babe Ruth. McGwire walked a record 162 times in 1998 and most of these were not intentional. Pitchers are simply terrified of McGwire. Sometimes a pitcher with two strikes and no balls would walk him simply because he was not going to give him a ball he could hit. When Babe Ruth came to the plate, a pitcher could walk him to get to Lou Gehrig, not an appetizing menu. When one of us (Watts) was pitching, I would have walked McGwire every time I faced him. Why fool around?

But we can also look at the ultimate aim of a baseball team, which is to score runs. As an admittedly unscientific measure, let us look at the ability of an individual to produce runs, which is the offensive strategy for winning games. In 1998, Tony Gwynn batted .321 and hit 16 home runs, while McGwire hit .299 with 70 home runs and Sosa hit .308 with 66 home runs. As a measure of effectively producing runs, we look at runs batted in (RBIs) plus runs scored (Rs) minus home runs (HRs). RBIs plus Rs needs to subtract HRs, because both RBIs and Rs include HRs. When you examine this number, you find that Gwynn produced 118 runs in 1998, while McGwire produced 207 and Sosa produced 226! Hitting home runs is not just a spectator event for fans; it wins ballgames. But as we pointed out in Chapter 1, one man does not make a winning team in the game of baseball. In the 1999 season, McGwire hit 65 homers and Sosa hit 63, but neither of their teams finished with a winning record.

There is an old story about a man who waited impatiently for his wife to get ready to go to a baseball game. When they finally arrived at the ballpark, she noted that the scoring line was no runs, no hits and no errors for both teams. "Look," she said, "nothing has happened. We haven't missed a thing." To baseball purists like us, plenty had happened. But today people want the excitement of home runs. Home runs bring more people to the ballpark. More power to them. And more power to Mark and Sammy and to the others who are increasing their home run production. The Babe increased interest in the game when he started hitting the long ball, and Mark and Sammy have done it again. Good.

Notes

1. John Thorn and Pete Palmer, *The Hidden Game of Baseball*, Doubleday, New York, 1984. Lots of interesting data and the authors' own methods of calculating who the best players were.

2. R. Cramer, "Average Batting Skill Through Major League History," *Baseball Research Journal*, 1980. One writer's opinion on how to compare different batters from different eras.

3. W. Rubinstein, "Commentary on 'Average Batting Skill Through Major League History,'" *Baseball Research Journal*, 1981. Interesting rebuttal to Cramer's method of comparing batters.

4. S. J. Gould, "Why No One Hits .400 Any More," *Discover*, vol. 7, August 1986. Delightful insights by a lover of baseball and science.

5. Joseph L. Reichler, ed., *The Baseball Encyclopedia*, 7th ed., Macmillan, New York, 1988. All the statistics you need and more.

Index

Index

Index

Index

simulation of, 190–191
validation of model for, 199–201
Robins, Benjamin, 47, 48
Robinson, Frank, 13
Robinson, Jackie, 8, 12
Rochester, New York, 3
roll (rotation of head), 175, 177n, 181
Rommel, Ed, 15
Rosen, Al, 111
rotation
of curve ball, 62
of various pitches, 59
see also axis of rotation
rotation rate
of batted-ball, 146, 147, 149–151, 164–165
of knuckleball, 91, 92
precollision, and collision results, 149–151
rotational motion of bat, 133–142
rounders, 1–4
Rover Boys, 20, 65
Rubenstein, W., 227–228
rule changes, 6–8, 231–233
Runge-Kutta formulation, 161
running, 230–231
Ruth, Babe, 14, 104–105, 111, 227, 244–248
Ryan, Nolan, 92
Ryan dead-ball, 103

saccadic eye movements, 171, 172, 177, 185–186, 193, 208, 223
St. Louis Cardinals, 109
San Francisco Giants, 116, 118–121, 123, 129, 141, 144
Sawyer, E., 82
scalar, 26
Schurr, Karl, 188
Science 82 magazine, 67
screwball, 15, 22, 58
scuff balls, 59, 62, 85–85, 90–91, 205
seams (of baseball), 60–61, 74, 75, 81–83
Selin, C., 92
separation point, boundary layer, 55, 56, 60, 62, 204, 205
Seymour, Harold, 3

shearing forces, 49, 54, 60
shine ball, 15, 232
Shirley, Bill, 92
sidearm curve, 69
SigmaPlot™, 131
Sikorsky, Igor, 67, 73, 76
Simple Model, 131–132
sinker, 15
sinusoid, 217
Sisler, George, 105, 243
slider, 59
sliding, 7
slippery elm, 62
Smith, Robert, 8, 20
smooth-pursuit eye movements, 172, 175, 177, 182, 193, 208, 216, 222–223
soaking, 4
softball, 4, 185
coefficient of restitution of, 120
and bat weights, 120, 122, 124, 125, 133
Sosa, Sammy, 6, 247, 248
Spalding, Albert G., 1, 2, 103
Spalding, J. Walter, 103
Spalding Bros., Inc., 103
Spalding League Ball, 103
Speaker, Tris, 243
speed
and bat-ball collisions, 106–109
of various pitches, 59
see also bat speed; batted-ball speed
speed estimating (of pitches), 197–198
spin (of ball), 6, 69
curve balls, 48, 89, 203–205
knuckleball, 61
of musket balls, 47, 48–49
and oblique collisions of ball and bat, 145
see also backspin; topspin
spin axis, 160
spin parameter, 79, 80, 156
spin rate
and temperature and humidity, 168–169
and trajectories, 162–164, 166–167
of various pitches, 59
spitball, 15, 59, 62, 231, 232

256